Illustrated Essentials of
MUSCULOSKELETAL ANATOMY

Second Edition

Illustrated Essentials of Musculoskeletal Anatomy

SECOND EDITON

KAY W. SIEG, M. ED., OTR
Associate Professor and Chairperson
Department of Occupational Therapy
College of Health Related Professions
University of Florida
Gainesville, Florida

SANDRA P. ADAMS, M.O.T., OTR
Assistant Professor
Department of Occupational Therapy
College of Health Related Professions
University of Florida
Gainesville, Florida

With contributions from the first edition
ANNA DEANE SCOTT, M.ED., OTR
Associate Professor
Department of Occupational Therapy
Sargent College of Allied Health Professions
Boston University
Boston, Massachusetts

MEGABOOKS

Gainesville, Florida 1985

Copyright© 1985 by Megabooks, Inc.

Second Edition

All rights reserved. This book is protected by copyright. No part of this book may be reproduced in any form or by an means, including photocopying, or utilized by any information storage and retrieval system without written permission from the copyright owner.

MEGABOOKS

Published by: Megabooks, Inc.
                4300 N.W. 23 Ave., Suite 192
                P.O. Box 1702
                Gainesville, FL 32602

Previous editions copyrighted in 1977.
Reprinted 1979.
Reprinted 1982.
Printed in the United States of America

Illustrators:

    Dan Conway
    Lewis Clark
    Brian Grunke
    Patricia Hobson
    Debra Neill-Mareci

Composition by Communication Graphics

Library of Congress Catalog Card Number 85-06-2295
ISBN 0-935157-00X

PREFACE TO SECOND EDITION

The students' enthusiasm for this simplified but comprehensive text on musculoskeletal anatomy was the impetus for our revising the first edition of this book. In this second edition we have included improved illustrations and additional content for clarification purposes. The most outstanding change is in the appearance of the book which is enhanced by a revised layout, a new cover design and printing process to improve its readability. It continues to provide concise but essential anatomical information in an easy-to-use reference.

PREFACE TO FIRST EDITION

This book provides clear, concise information on musculoskeletal anatomy for both the individual who is seeking a review reference as well as for the beginning anatomy student. Many texts offer detailed anatomical information; however, for the novice anatomy student, the essentials are sometimes difficult to abstract. The authors, thus, created this manual to indicate the basic anatomical information which we wanted the student therapist and health professional to learn.

Since this manual was designed to present the essentials of musculoskeletal anatomy, the scope of the information included reflects this intent. A comprehensive anatomy text should be consulted for composite pictures depicting the relative position of the muscles in superficial and deep dissection, blood supply, intricate musculature of the head, body wall or genitalia, or details of bones, joints and ligaments. Such information is readily found in any of the anatomy books listed in the references at the end of this manual.

In this manual, separate illustrations and descriptions of individual skeletal muscles, bones and nerves of the upper and lower extremities are provided. Also included are: muscle groups which perform the motion of the joints; major muscle groups of the head, neck and trunk; and locations for muscle and bone palpation on the living body. Therefore, the book is appropriate for anyone seeking simplified information about individual muscle location, innervation and function, as well as the muscle groups which provide the elements of motion needed to perform everyday human activities.

TABLE OF CONTENTS

INSTRUCTIONS FOR USE OF MANUAL ... xi

UPPER EXTREMITY

Bones ... 1
- skeleton — 1
- skull — 2
- vertebral column — 3
- sternum, clavicle, rib — 4
- scapula — 5
- humerus — 6
- radius-ulna — 7
- hand — 8
- palpation — 9, 10

Muscles ... 11
- shoulder — 11
- arm — 22
- forearm — 28
- hand — 46

Motions ... 58
- scapula — 58
- humerus — 61
- elbow — 67
- forearm — 68
- wrist — 69
- digits — 71

Nerves ... 73
- brachial plexus — 73
- axillary nerve — 74
- musculocutaneous nerve — 75
- radial nerve — 76
- median nerve — 77
- ulnar nerve — 78
- cutaneous innervation — 79

LOWER EXTREMITY

Bones ... 80
- pelvis — 80
- femur — 81
- tibia-fibula — 82
- foot — 83
- palpations — 84, 85

Muscles ... 86
- hip — 86
- thigh — 92
- leg — 102
- foot — 115

Motions ... 119
- hip — 119
- knee — 122
- ankle — 124

Nerves ... 126
- lumbosacral plexus — 126
- obturator nerve — 127
- femoral nerve — 128
- sciatic nerve — 129
- peroneal nerve — 130
- tibial nerve — 131
- cutaneous innervation — 132

HEAD, NECK, TRUNK MUSCLES

- **Extraocular** ... 133
- **Mastication, Lip and Jaw** ... 134, 135
- **Deglutition** ... 136
- **Neck** ... 137
- **Abdomen** ... 139
- **Respiration** ... 143
- **Back** ... 146

REFERENCES ... 151

INDEX ... 152

INSTRUCTIONS FOR USE OF MANUAL

1. **Origin, insertion, action, nerve**

    The main origin, insertion, action, nerve and spinal segments for each muscle are presented as follows:

    O. – origin
    I. – insertion
    A. – action
    N. – nerve (spinal segments)

2. **Palpation**

    Muscle palpations are signified by P. and are briefly described for most muscles. Palpations should be done by resisting the action of the muscle and palpating the muscle belly and/or tendon, when specified. For example:

    P. – Anterior surface of humerus (indicates palpation site of biceps brachii muscle belly)

    P. – Tendon palpated in antecubital fossa (indicates palpation site of tendon of insertion of biceps brachii)

3. **Actions**

    Muscles that serve as prime movers for a particular joint action are listed first. Muscles which assist in that action are in italics. For example, for Flexors of the Elbow:

    BICEPS BRACHII (prime mover)
    Pronator teres (assists)

4. **Nerves**

    There are separate nerve drawings for motor and cutaneous distributions.

5. **Abbreviations**

    CM – carpometacarpal
    DIP – distal interphalangeal
    IP – interphalangeal
    MP – metacarpophalangeal in hand, metatarsophalangeal in foot
    PIP – proximal interphalangeal

6. **Comments**

    Memorization tips and/or notes of interest concerning the muscle are provided for each individual muscle.

Skeleton

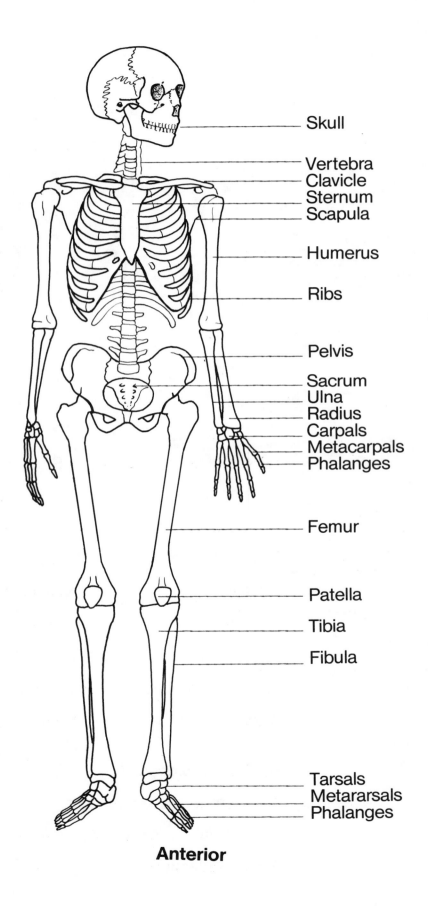

Anterior

Skull

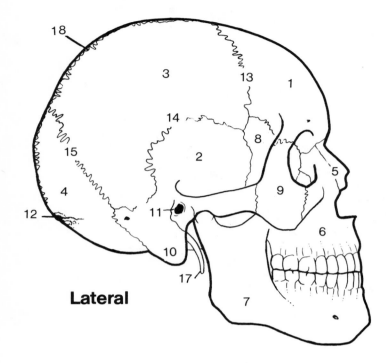

Lateral

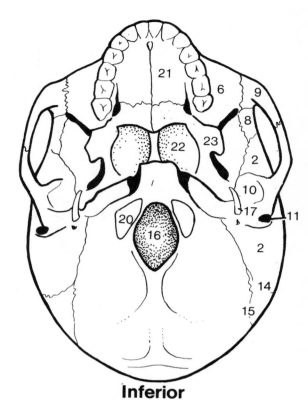

Inferior

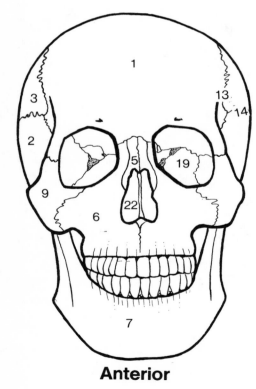

Anterior

1. Frontal bone
2. Temporal bone
3. Parietal bone
4. Occipital bone
5. Nasal bone
6. Maxilla
7. Mandible
8. Sphenoid bone
9. Zygomatic bone and arch
10. Mastoid process
11. External auditory meatus
12. External occipital protuberance
13. Coronal suture
14. Squamous suture
15. Lamboidal suture
16. Foramen magnum
17. Styloid process
18. Sagittal suture
19. Orbit
20. Occipital condyle
21. Palatine process of maxilla
22. Nasal aperture
23. Pterygoid plates

Vertebral Column

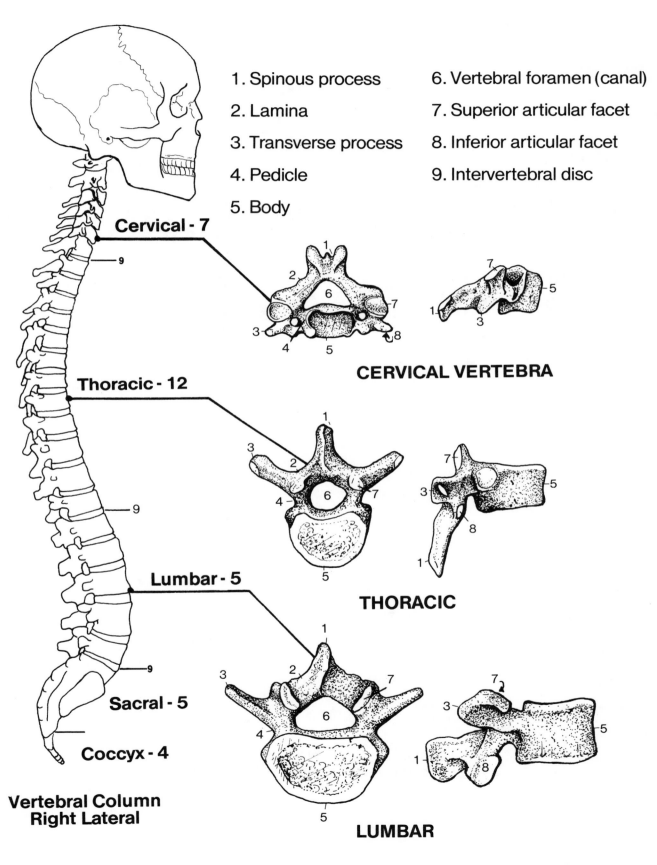

Sternum, Rib and Right Clavicle

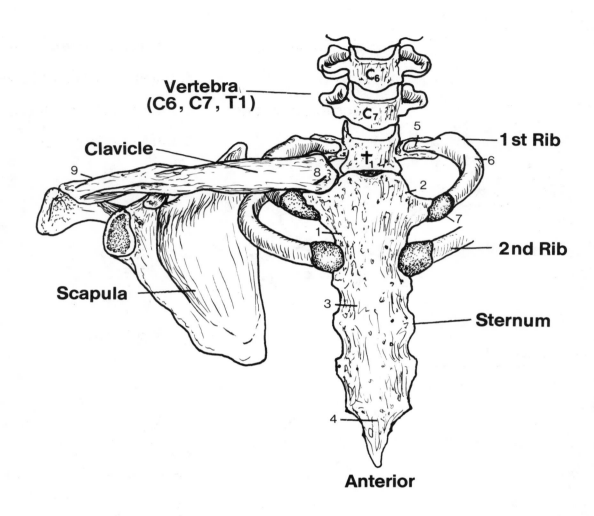

Anterior

Sternum
1. Manubrium
2. Clavicular notch
3. Body
4. Xiphoid process

Rib
5. Head
6. Body
7. Costal cartilage

Clavicle
8. Sternal end
9. Acromial end

Vertebra and scapula are included here to depict anatomical relationships and are described in detail on pages 3 and 5, respectively.

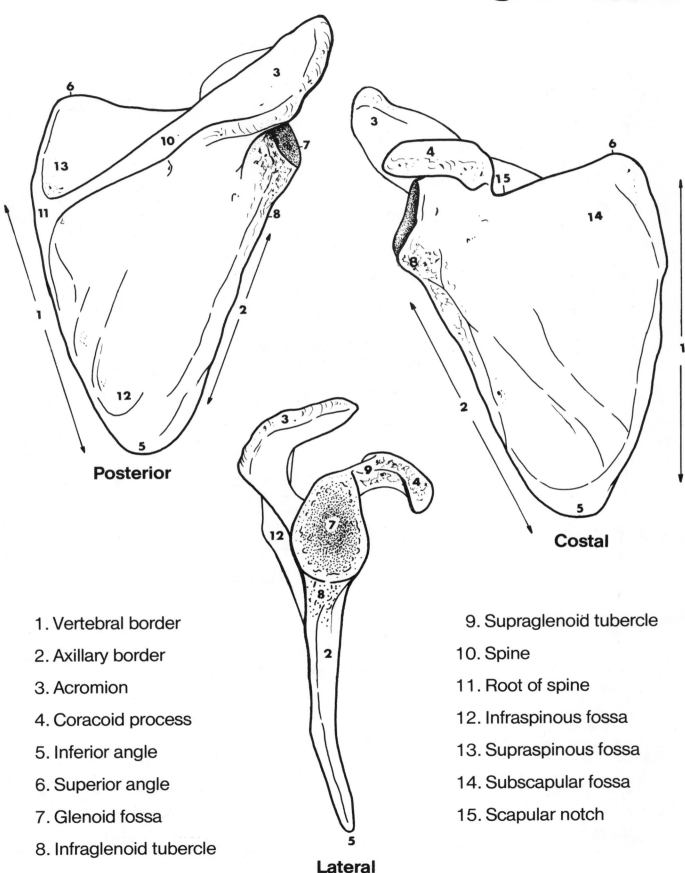

Right Humerus

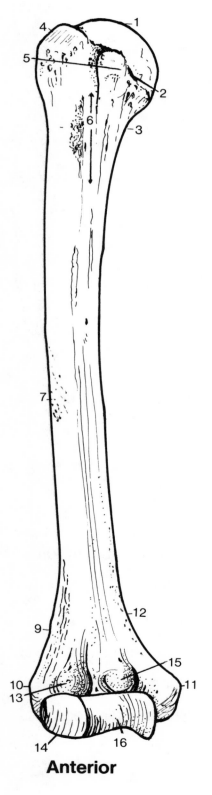

1. Head
2. Anatomical neck
3. Surgical neck
4. Greater tubercle
5. Lesser tubercle
6. Intertubercular groove (bicipital groove)
7. Deltoid tuberosity
8. Groove for radial nerve (spiral groove)
9. Lateral supracondylar ridge
10. Lateral epicondyle
11. Medial epicondyle
12. Medial supracondylar ridge
13. Radial fossa
14. Capitulum
15. Coronoid fossa
16. Trochlea
17. Olecranon fossa

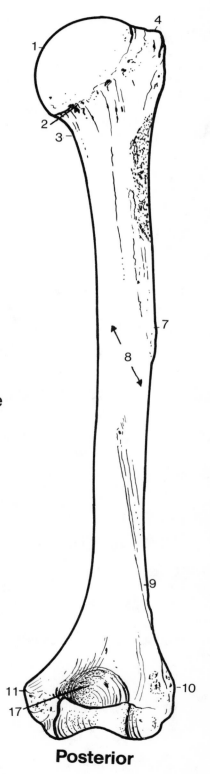

Anterior

Posterior

Right Radius and Ulna

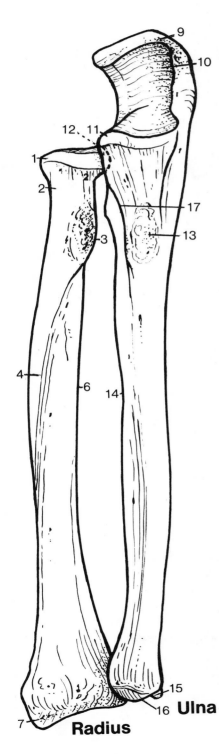

Anterior

Radius
1. Head
2. Neck
3. Tuberosity
4. Anterior oblique line
5. Posterior oblique line
6. Interosseous border
7. Styloid process
8. Dorsal tubercle

Ulna
9. Olecranon process
10. Trochlear notch (semilunar notch)
11. Coronoid process
12. Radial notch
13. Tuberosity
14. Interosseous border
15. Styloid process
16. Head
17. Supinator crest

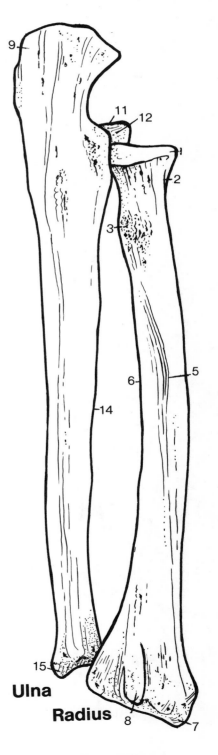

Posterior

7

Right Hand

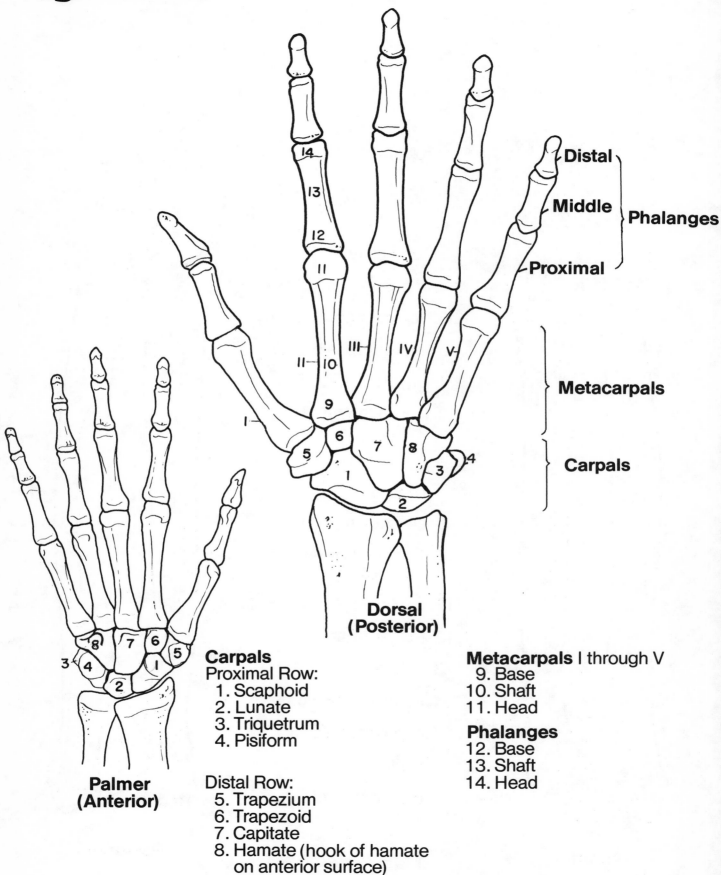

Carpals
Proximal Row:
1. Scaphoid
2. Lunate
3. Triquetrum
4. Pisiform

Distal Row:
5. Trapezium
6. Trapezoid
7. Capitate
8. Hamate (hook of hamate on anterior surface)

Metacarpals I through V
9. Base
10. Shaft
11. Head

Phalanges
12. Base
13. Shaft
14. Head

Palpation of Boney Landmarks – Upper Extremity and Head

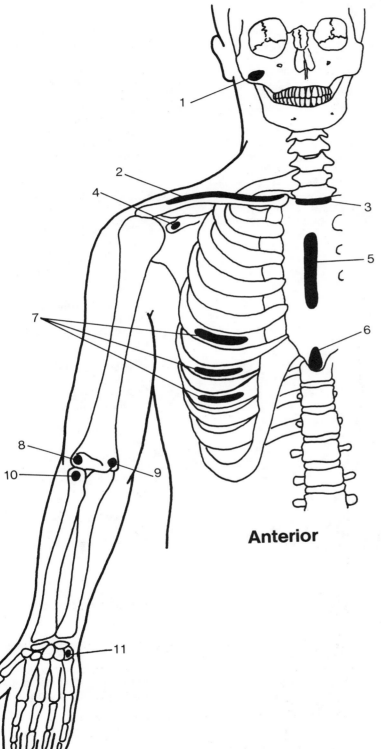

Anterior

1. Zygomatic bone – "cheek bone"

2. Clavicle – "collar bone" – From its medial end at sternoclavicular joint to its lateral end at acromioclavicular joint

3. Sternal notch – At superior end of manubrium between medial ends of the two clavicles

4. Coracoid process of scapula – Inferior to lateral part of clavicle

5. Sternum – "breast plate" – Between ribs

6. Xiphoid process – "Sword-like," pointed process at inferior end of sternum

7. Ribs – At angle of rib on lateral thorax

8. Lateral epicondyle of humerus – Lateral bump on distal end of humerus

9. Medial epicondyle of humerus – Medial bump on distal end of humerus

10. Head of radius – Boney knob just distal to lateral epicondyle of humerus (with elbow extended, it can be felt rolling during supination and pronation)

11. Pisiform – Medial carpal bone at distal skin crease on anterior wrist

Palpation of Boney Landmarks – Upper Extremity and Head

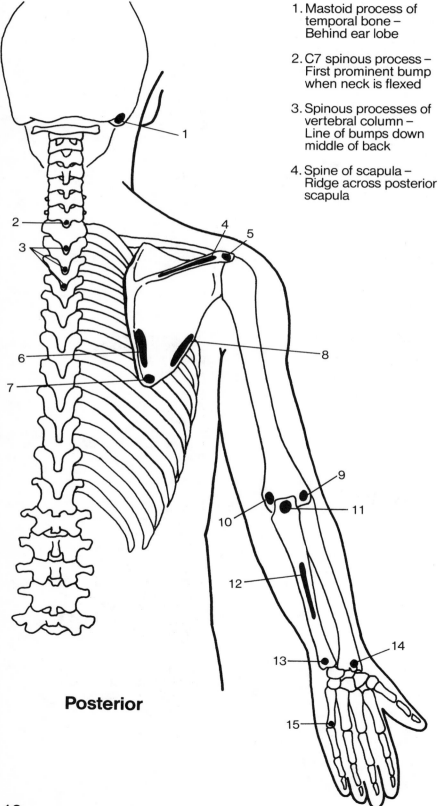

1. Mastoid process of temporal bone – Behind ear lobe
2. C7 spinous process – First prominent bump when neck is flexed
3. Spinous processes of vertebral column – Line of bumps down middle of back
4. Spine of scapula – Ridge across posterior scapula
5. Acromion process of scapula – Lateral end of spine of scapula; forms top of shoulder
6. Vertebral border of scapula – Medial border that runs parallel to vertebral column
7. Inferior angle at scapula – Lower angle
8. Axillary border of scapula – Lateral border from inferior angle to axilla (armpit)
9. Lateral epicondyle of humerus – Lateral bump on distal end of humerus
10. Medial epicondyle of humerus – Medial bump on distal end of humerus
11. Olecranon process of ulna – Point of elbow on posterior surface between lateral and medial epicondyles
12. Posterior border of ulna – Sharp dorsal margin of ulna; palpated entire length
13. Styloid process of ulna – A prominent knob on distal end of posterior ulna
14. Styloid process of radius – Distal end of lateral aspect of radius (it is slightly more distal than styloid process of ulna)
15. Metacarpal heads – "Knuckles" at MP joints

Posterior

Trapezius

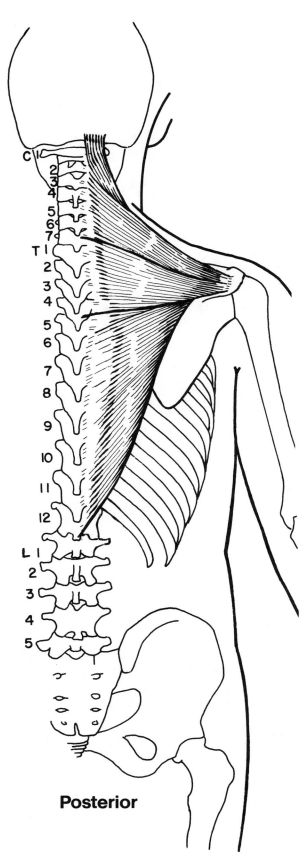

O. Occiput, ligamentum nuchae
C7-T12 (spinous processes)

I. Upper: lateral clavicle, acromion
Middle: spine of scapula
Lower: root of spine of scapula

A. Upper: elevation, upward rotation of scapula
Middle: retraction of scapula
Lower: depression, upward rotation of scapula

N. Accessory nerve (Cranial Nerve XI) and branches of C3, 4

P. Upper: between base of skull and lateral third of clavicle

Middle: between spinous processes of T1-T5 and vertebral border of scapula

Lower: between spinous processes of T6-T12 and root of spine

This is a triangular shaped muscle which, when paired, forms a trapezium.

Latissimus Dorsi

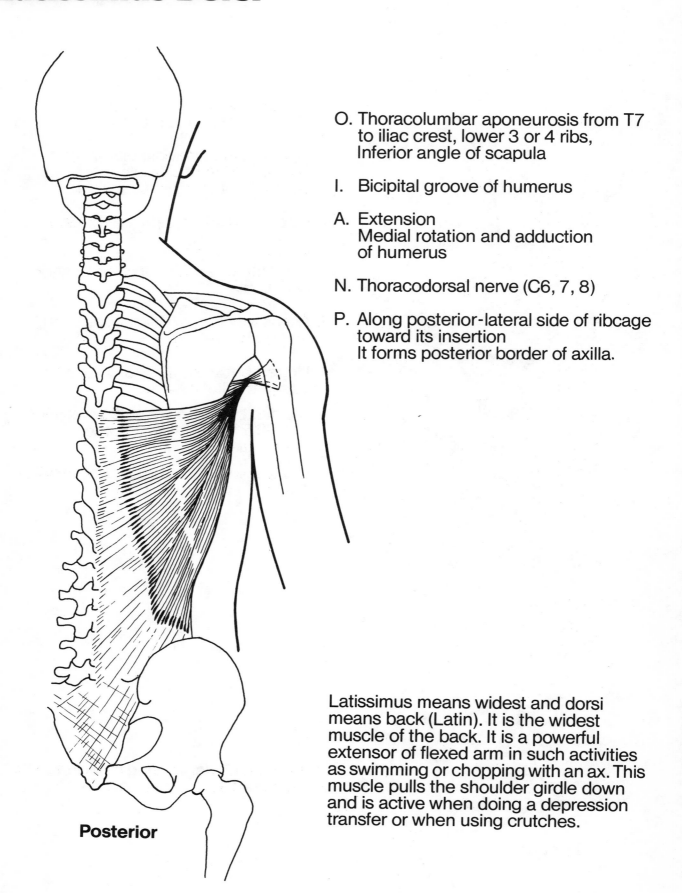

O. Thoracolumbar aponeurosis from T7 to iliac crest, lower 3 or 4 ribs, Inferior angle of scapula

I. Bicipital groove of humerus

A. Extension
Medial rotation and adduction of humerus

N. Thoracodorsal nerve (C6, 7, 8)

P. Along posterior-lateral side of ribcage toward its insertion
It forms posterior border of axilla.

Posterior

Latissimus means widest and dorsi means back (Latin). It is the widest muscle of the back. It is a powerful extensor of flexed arm in such activities as swimming or chopping with an ax. This muscle pulls the shoulder girdle down and is active when doing a depression transfer or when using crutches.

Teres Major

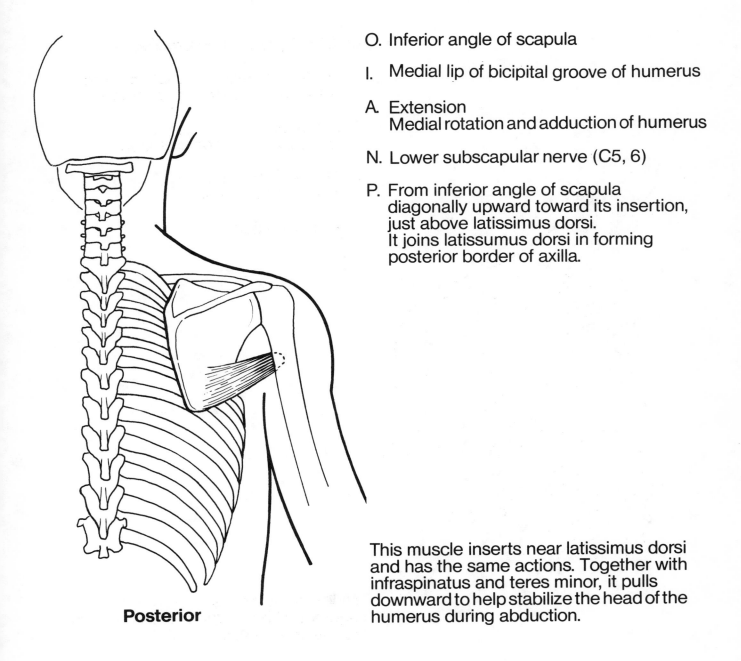

O. Inferior angle of scapula

I. Medial lip of bicipital groove of humerus

A. Extension
Medial rotation and adduction of humerus

N. Lower subscapular nerve (C5, 6)

P. From inferior angle of scapula diagonally upward toward its insertion, just above latissimus dorsi.
It joins latissumus dorsi in forming posterior border of axilla.

Posterior

This muscle inserts near latissimus dorsi and has the same actions. Together with infraspinatus and teres minor, it pulls downward to help stabilize the head of the humerus during abduction.

Levator Scapula

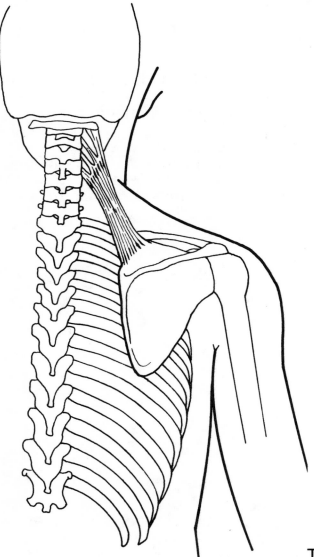

O. C1 - 4 (transverse processes)

I. Vertebral border of scapula from superior angle to root of spine

A. Elevation
Downward rotation of scapula

N. Dorsal scapular nerve (C5) and branches of C3, 4

P. Cannot palpate

The levator is an elevator (L - evator). Together with the upper trapezius, this muscle shrugs your shoulders.

Rhomboids: Major and Minor

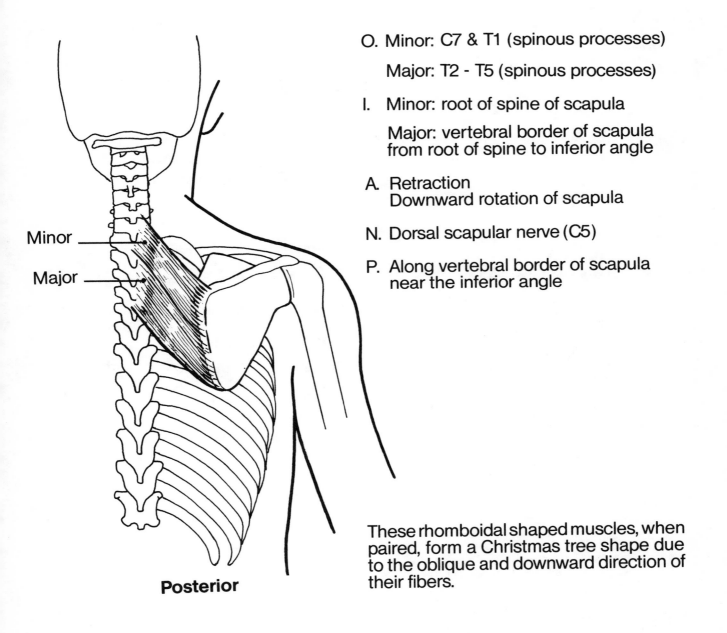

O. Minor: C7 & T1 (spinous processes)
   Major: T2 - T5 (spinous processes)

I. Minor: root of spine of scapula
   Major: vertebral border of scapula from root of spine to inferior angle

A. Retraction
   Downward rotation of scapula

N. Dorsal scapular nerve (C5)

P. Along vertebral border of scapula near the inferior angle

These rhomboidal shaped muscles, when paired, form a Christmas tree shape due to the oblique and downward direction of their fibers.

Deltoids

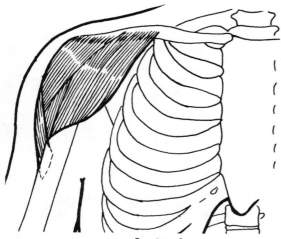

Anterior
Anterior Deltoid

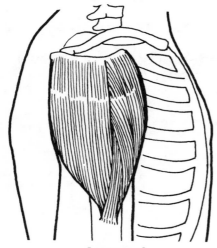

Lateral
Middle Deltoid

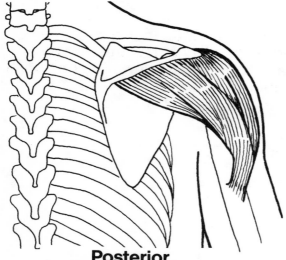

Posterior
Posterior Deltoid

O. Anterior: lateral third of clavicle

Middle: lateral acromion

Posterior: spine of scapula

I. Deltoid tuberosity of humerus

A. Anterior: flexion, horizontal adduction, medial rotation of humerus

Middle: abduction of humerus to 90°

Posterior: extension, horizontal abduction, lateral rotation of humerus

N. Axillary nerve (Circumflex) (C5, 6)

P. Anterior: anterior-medial surface of upper arm below acromion process, just anterior to glenohumeral joint

Middle: below acromion on lateral surface of upper arm, just lateral to glenohumeral joint

Posterior: posterior-lateral surface of upper arm, just posterior to glenohumeral joint

This muscle forms the rounded muscle bulk over the shoulder joint. It is a strong abductor, but cannot initiate that movement because the angle of pull is parallel to the humerus when the arm is by the side. See supraspinatus.

Coracobrachialis

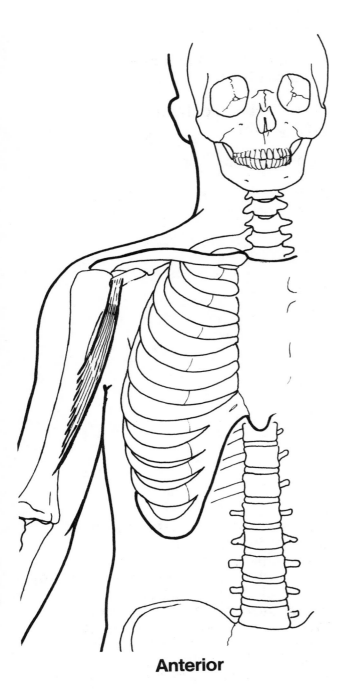

Anterior

O. Coracoid process of scapula

I. Middle of medial border of humeral shaft

A. Flexion
   Adduction of humerus

N. Musculocutaneous nerve (C5, 6, 7)

P. Difficult to palpate

This muscle is named for its origin and insertion.

Supraspinatus

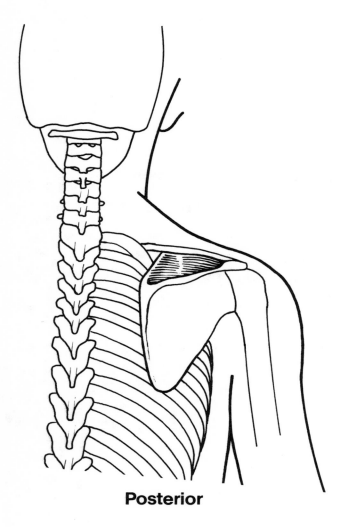

Posterior

O. Supraspinous fossa of scapula

I. Greater tubercle of humerus (superior facet)

A. Stabilize head of humerus to initiate abduction

N. Suprascapular nerve (C5, 6)

P. Above spine of scapula (if trapezius is relaxed by bringing ear to shoulder)

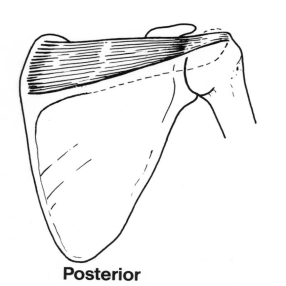

Posterior

The supraspinatus, infraspinatus and teres minor insert, respectively, on the superior, middle and inferior facets of the greater tubercle of the humerus and thus can be recalled by the first letter of their names as the "SIT" muscles, supraspinatus being the S of the SIT muscles. This muscle pulls the head of the humerus into the glenoid fossa to initiate abduction and thus provides an effective angle of pull for the deltoid. It reinforces the capsule of the shoulder joint and is one of the "rotator cuff" muscles.

Infraspinatus

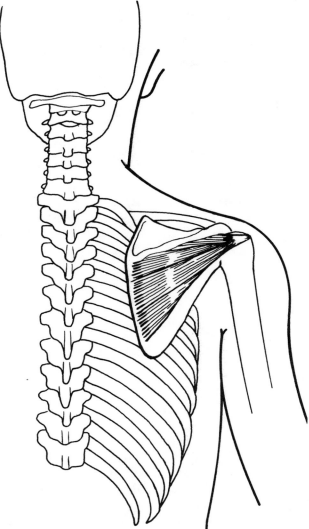

Posterior

O. Infraspinous fossa of scapula

I. Greater tubercle of humerus (middle facet)

A. Lateral rotation
Extension of humerus

N. Suprascapular nerve (C5, 6)

P. Below spine of scapula

Infraspinatus is the I of the SIT muscles (see supraspinatus and teres minor). It reinforces the capsule of the shoulder joint and is one of the "rotator cuff" muscles. This muscle together with teres minor and teres major also helps stabilize the head of the humerus during abduction.

Teres Minor

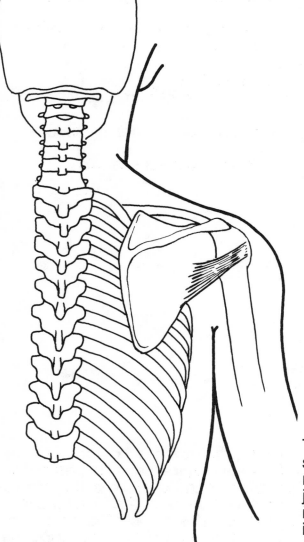

Posterior

O. Upper axillary border of scapula

I. Greater tubercle of humerus (inferior facet)

A. Lateral rotation
   Extension of humerus

N. Axillary nerve (circumflex) (C5, 6)

P. Between posterior deltoid and axillary border of scapula; located superior to teres major and inferior to infraspinatus

Teres minor is the T of the SIT muscles (see supraspinatus and infraspinatus). It reinforces the capsule of the shoulder joint and is one of the "rotator cuff" muscles. This muscle together with infraspinatus and teres major also helps stabilize the head of the humerus during abduction.

Subscapularis

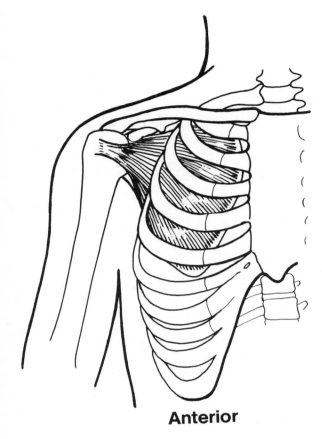

Anterior

O. Subscapular fossa of scapula

I. Lesser tubercle of humerus

A. Medial rotation of humerus

N. Upper and lower subscapular nerve (C5, 6)

P. Belly is difficult to palpate

Tendon palpated deep in middle of axilla

Anterior

Subscapularis, supraspinatus, infraspinatus, and teres minor form the "rotator cuff" to guard the glenohumeral joint. Subscapularis inserts on the lesser tubercle and the other 3 (SIT muscles) insert on the greater tubercle.

Pectoralis major

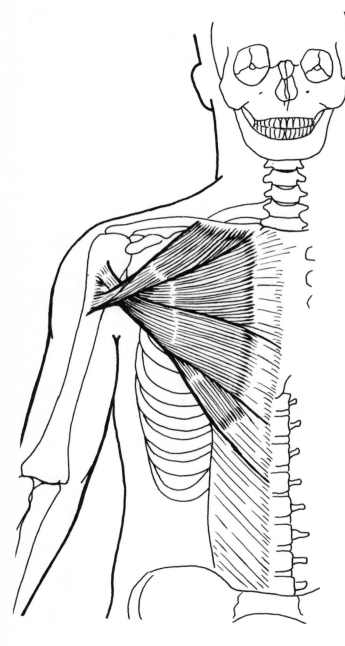

Anterior

O. Clavicular head: medial half of clavicle
   Sternal head: sternum, cartilages of upper 6 ribs

I. Lateral lip bicipital groove of humerus

A. Adduction, horizontal adduction and medial rotation of humerus

   Clavicular head: flexion of humerus

   Sternal head: extension of humerus from a flexed position

N. Clavicular head: lateral pectoral nerve (C5, 6, 7)
   Sternal head: medial pectoral nerve (C8, T1)

P. Along anterior border of axilla

This muscle forms the anterior wall of the axilla.

Pectoralis Minor and Subclavius

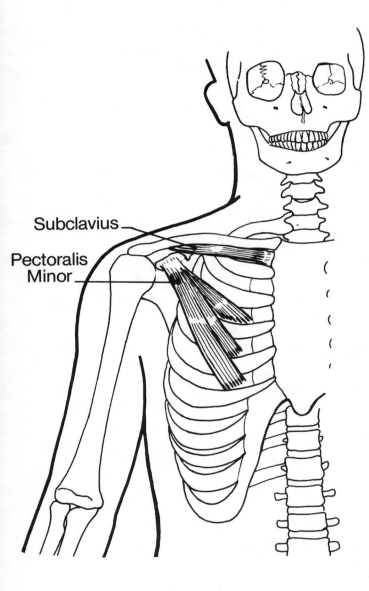

Anterior

Pectoralis Minor

O. Anterior 3, 4, 5 ribs

I. Coracoid process of scapula

A. Protraction, depression, downward rotation of scapula

N. Medial pectoral nerve (C8, T1)

P. Difficult to palpate

Pectoralis minor forms a bridge over the brachial plexus and vessels of the arm.

Subclavius

O. 1st rib

I. Inferior shaft of clavicle

A. Stabilizes clavicle

N. Nerve to subclavius (C5, 6)

P. Cannot palpate

Subclavius prevents extreme elevation and protraction of the clavicle.

Serratus Anterior

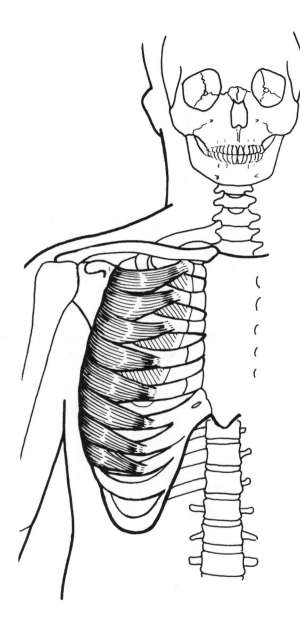

Anterior

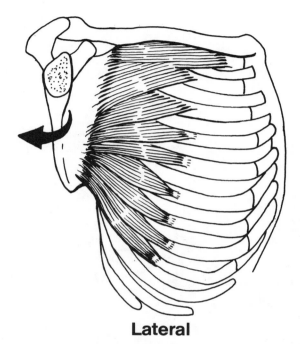

Lateral

O. Outer surface of upper 8 ribs (by finger-like slips)

I. Vertebral border of scapula

A. Protraction, upward rotation of scapula, stabilizes scapula against chest wall

N. Long thoracic nerve (C5, 6, 7)

P. Lateral-anterior surface of ribs, below axilla

This muscles derives its name from the serrated origin. The lower fibers of the origin interdigitate with the external oblique. It is the strongest protractor of the scapula and holds the scapula against the chest wall to provide a fixed origin for muscles acting on the humerus. Weakness causes "winged scapula."

Biceps Brachii

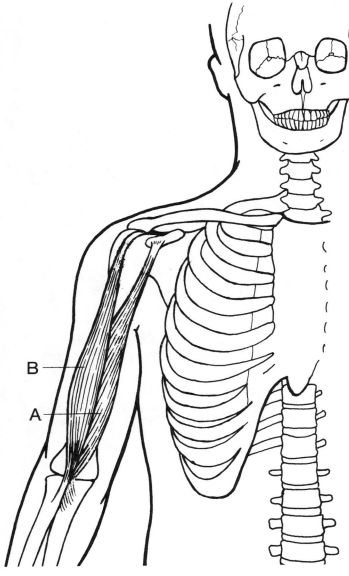

Anterior

O. A. Short head: coracoid process of scapula
B. Long head: supraglenoid tubercle of scapula

I. Tuberosity of radius

A. Flexion of elbow
Supination of forearm
Short head – flexion of humerus

N. Musculocutaneous nerve (C5, 7)

P. Anterior surface of humerus

Tendon palpated in anticubital fossa

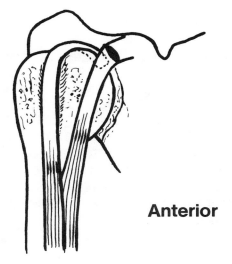

Anterior

Biceps means two heads and brachium means arm (Latin). The long head travels through the capsule of the shoulder joint and the bicipital groove. The short head originates on the coracoid process along with the origin of the coracobrachialis and the insertion of the pectoralis minor.

Brachialis

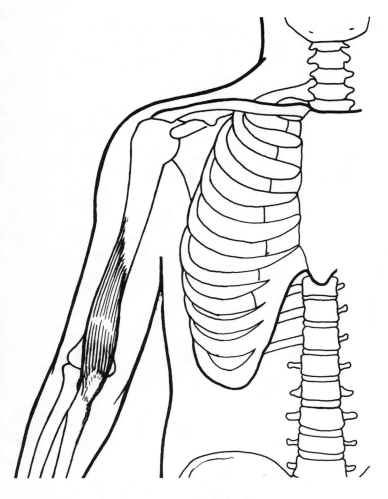

Anterior

O. Lower half of anterior shaft of humerus

I. Tuberosity of ulna

A. Flexion of elbow

N. Musculocutaneous nerve (C5, 6, 7) (sometimes branches from radial and median nerves)

P. Medial to biceps on lower anterior humerus (relax biceps by pronating forearm)

Tendon palpated deep in anticubital fossa just medial to biceps tendon

This muscle is the strongest elbow flexor.

Triceps Brachii and Anconeus

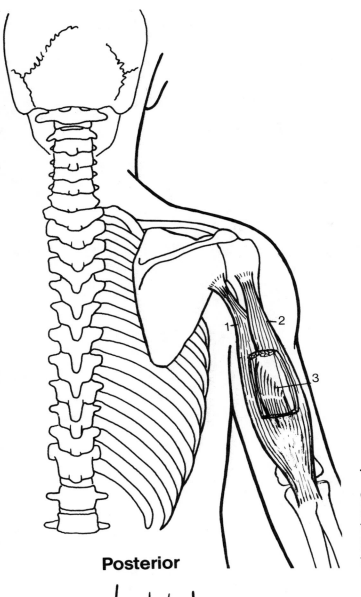

Posterior

Posterior

Triceps Brachii

O. 1. Long head: infraglenoid tubercle of scapula

   2. Lateral head: posterior humerus above spiral groove

   3. Medial head: posterior humerus below spiral groove

I. Olecranon process of ulna

A. Extension of elbow
Long head: extension of humerus

N. Radial nerve (C7, 8)

P. Posterior and lateral surface of humerus

Triceps means three heads (Latin). It is the only muscle on the posterior arm. The long head of the triceps originates from the infraglenoid tubercle, whereas the long head of the biceps originates from the supraglenoid tubercle.

Anconeus

O. Lateral epicondyle of humerus

I. Olecranon process of ulna

A. Extension of elbow

N. Radial nerve (C7, 8)

P. Below elbow joint on posterior-lateral surface of upper forearm between olecranon and lateral epicondyle

Brachioradialis

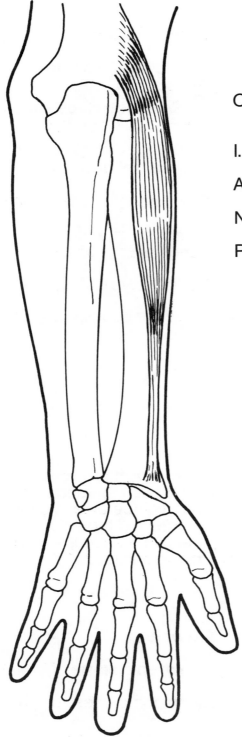

Posterior

O. Lateral supracondylar ridge of humerus

I. Styloid process of radius

A. Flexion of elbow in neutral position

N. Radial nerve (C5, 6)

P. Lateral surface of radius on upper forearm

This muscle derives its name from its origin and insertion. It flexes the elbow with the forearm in mid-positon, whereas the biceps flexes the elbow in supination and the brachialis flexes the elbow in all positions. Brachioradialis functions when the elbow flexes against resistance or rapidly. Acting alone, this muscle would tend to bring the forearm to mid-position from either a supinated or pronated position as it flexes the elbow.

Extensor Carpi Radialis Longus

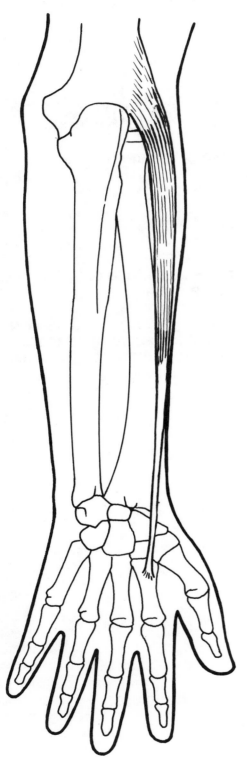

O. Lateral supracondylar ridge
   Lateral epicondyle of humerus

I. Base of 2nd metacarpal

A. Extension
   Abduction of wrist (radial deviation)

N. Radial nerve (C6, 7)

P. Dorsal proximal foreman, adjacent to brachioradialis

Tendon palpated on dorsal surface of wrist at base of 2nd metacarpal

Acting alone, this muscle extends the wrist in a radial direction.

Posterior

Extensor Carpi Radialis Brevis

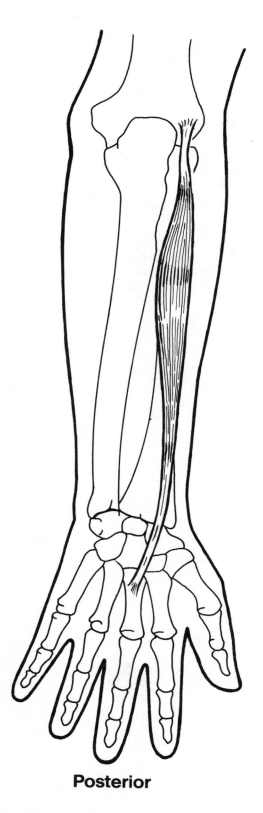

Posterior

O. Lateral epicondyle of humerus (common extensor tendon)

I. Base of 3rd metacarpal

A. Extension of wrist

N. Radial nerve (C6, 7)

P. Distal to extensor carpi radialis longus on dorsal surface of forearm; difficult to differentiate

Tendon palpated on dorsal surface of wrist at base of third metacarpal, medial to extensor carpi radialis longus

Acting alone, this muscle extends the wrist in the midline.

Extensor Carpi Ulnaris

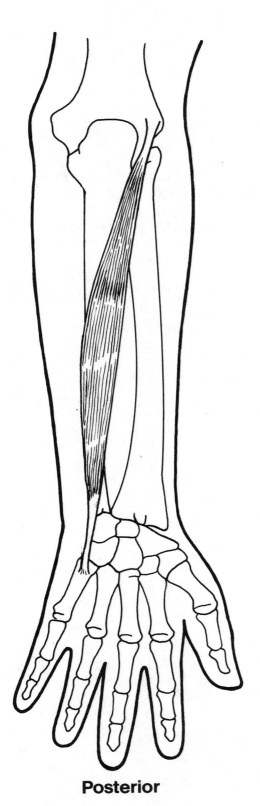

Posterior

O. Lateral epicondyle of humerus (common extensor tendon)
Posterior proximal ulna

I. Base of 5th metacarpal

A. Extension
Adduction of wrist (ulnar deviation)

N. Radial nerve (C6, 7, 8)

P. Along ulnar border of dorsal forearm

Tendon palpated on dorsal surface of wrist on ulnar side of carpal bones

Acting alone, this muscle extends the wrist in an ulnar direction. The tendon runs through a groove between the head of the ulna and the styloid process of the ulna.

Extensor Digitorum

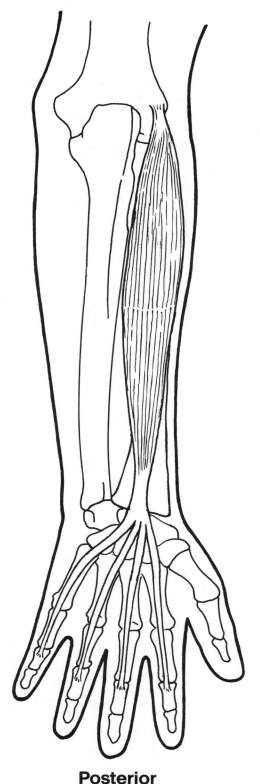

Posterior

O. Lateral epicondyle of humerus (common extensor tendon)

I. Base of middle phalanges of 4 fingers (dorsal surface)
Extensor expansion of 4 fingers

A. Extension of MP joints of 4 fingers

N. Radial nerve (C6, 7, 8)

P. Middle of dorsal forearm, distal and lateral to wrist extensor muscle bellies
Tendons palpated on dorsum of hand

This muscle divides into the four prominent tendons on the dorsum of the hand. The tendons run down to the base of the middle phalanges where they flatten out and spread into the extensor expansion to each finger. The muscle itself cannot extend the distal phalanges. Extension of the IP joints is due to the lumbricals and interossei. (See lumbricals and interossei.)

Extensor Digiti Minimi

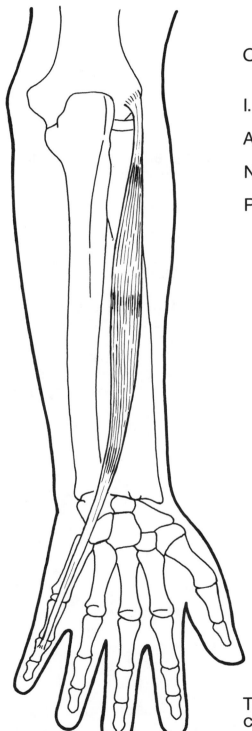

O. Lateral epicondyle of humerus (common extensor tendon)

I. Extensor expansion of little finger

A. Extension of little finger at MP joint

N. Radial nerve (C6, 7, 8)

P. Cannot palpate

This muscle joins the extensor digitorum communis tendon in the extensor expansion.

Posterior

Extensor Indicis

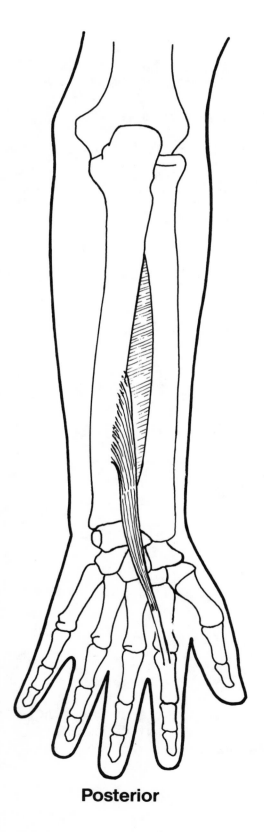

Posterior

O. Posterior ulna and interosseous membrane

I. Extensor expansion of index finger

A. Extension of index finger at MP joint

N. Radial nerve (C6, 7, 8)

P. Mid to distal forearm on dorsal surface between radius and ulna

Tendon palpated in area of MP joint on ulnar side of extensor digitorum tendon to index finger (dorsal surface)

There are 4 deep muscles originating on the posterior radius, ulna and interosseous membrane. From most proximal to distal in origin, they are abductor pollicis longus, extensor pollicis brevis, extensor pollicis longus and extensor indicis.

Supinator

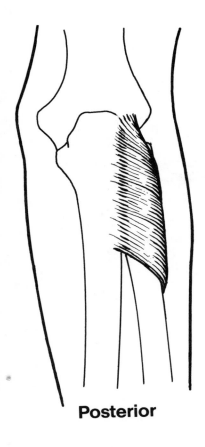

Posterior

O. Below the radial notch on posterior ulna and posterior capsule

I. Between anterior and posterior oblique lines of proximal radius on anterior surface

A. Supination of forearm

N. Radial nerve (C6)

P. Cannot palpate, but can test for for function

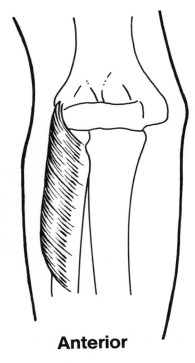

Anterior

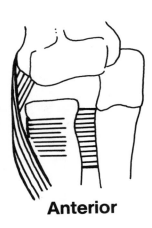

Anterior

For slow, non-resisted supination and for supination with the elbow extended, the supinator is sufficient. When speed or resistance is required, the stronger supinator, biceps brachii, is recruited. The radial nerve passes through this muscle.

Abductor Pollicis Longus

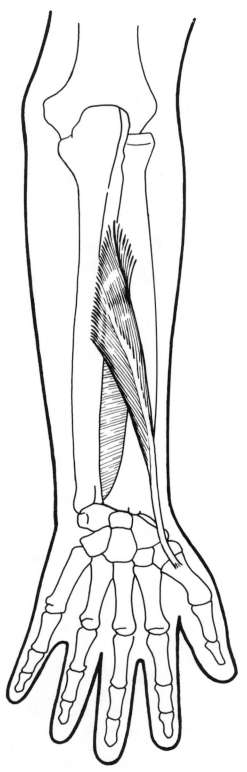

Posterior

O. Posterior radius, ulna and interosseous membrane

I. Base of 1st metacarpal

A. Abduction of thumb at CM joint

N. Radial nerve (C6, 7)

P. Tendon palpated at wrist joint on radial side of base of 1st metacarpal

Acting alone, this muscle abducts the thumb in a radial direction. By continued action it can flex and abduct the wrist.

Extensor Pollicis Brevis

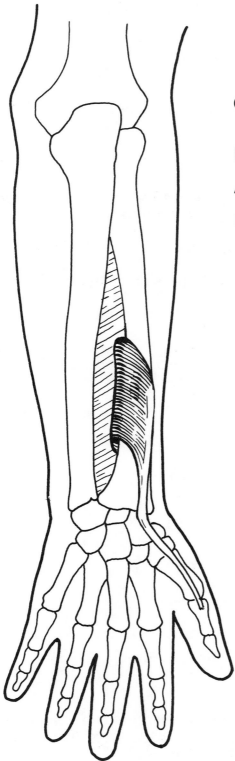

Posterior

O. Posterior radius
Interosseous membrane

I. Base of proximal phalanx of thumb

A. Extension of thumb at MP joint

N. Radial nerve (C6, 7)

P. Tendon palpated on radial side of "anatomical snuffbox," adjacent to abductor pollicis longus tendon

This muscle forms the lateral border of the "anatomical snuffbox," while extensor pollicis longus forms the medial border. The snuffbox is a depression on the dorsum of the 1st metacarpal reputedly used to hold snuff for sniffing by genteel persons in the past century.

Extensor Pollicis Longus

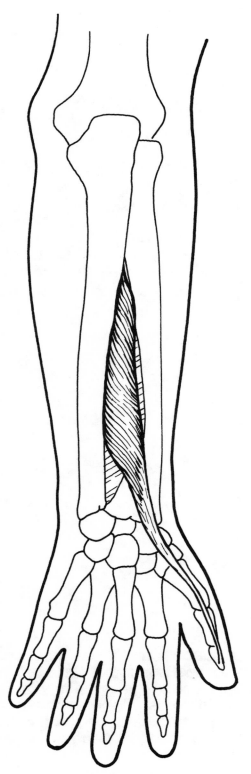

O. Posterior ulna and interosseous membrane

I. Base of distal phalanx of thumb

A. Extension of thumb at IP joint

N. Radial nerve (C6, 7, 8)

P. Tendon palpated at ulnar side of "anatomical snuffbox" and also on dorsal proximal phalanx

This muscle forms the medial border of the "anatomical snuffbox" (see extensor pollicis brevis).

Posterior

Pronator Teres and Pronator Quadratus

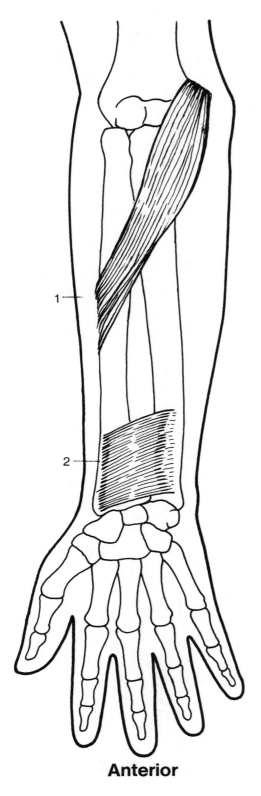

Anterior

1. **Pronator Teres**

   O. Above the medial epicondyle of humerus, coronoid process of ulna

   I. Middle of lateral shaft of radius

   A. Pronation of forearm
   Assists in flexion of elbow

   N. Median nerve (C6, 7)

   P. Medial side of anterior surface of proximal forearm, just medial to biceps insertion

   Pronator teres is prominent in resisted pronation. Its fibers run from origin to insertion diagonally across the proximal half of the anterior forearm.

2. **Pronator Quadratus**

   O. Distal fourth of anterior ulna

   I. Distal fourth of anterior radius

   A. Pronates forearm

   N. Median nerve (C8, T1)

   P. Cannot palpate

   Pronator quadratus functions in resisted and non-resisted pronation. It has more fibers in cross section than pronator teres.

39

Flexor Carpi Ulnaris

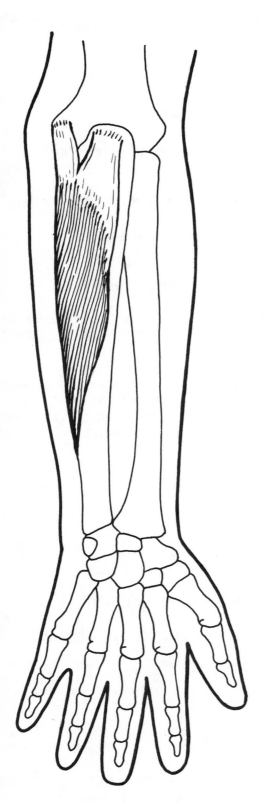

O. Medial epicondyle of humerus
   Proximal posterior ulna

I. Pisiform, hamate, and base of 5th metacarpal

A. Flexion
   Adduction of wrist

N. Ulnar nerve (C8, T1)

P. Tendon palpated on anterior surface of wrist, proximal to pisiform bone

Acting alone, this muscle flexes the wrist in an ulnar direction. Flexor carpi ulnaris and extensor carpi ulnaris function together to adduct the wrist. The ulnar nerve passes through this muscle.

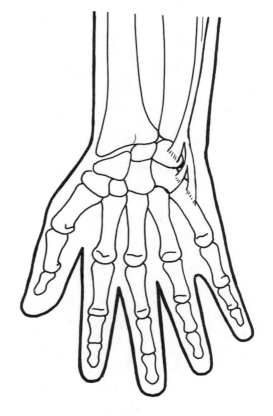

Posterior   **Anterior**

Palmaris Longus

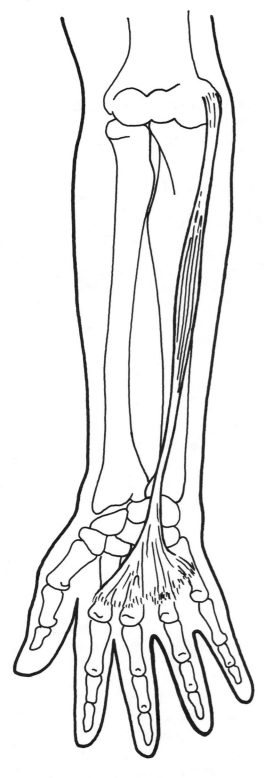

Anterior

O. Medial epicondyle of humerus

I. Palmar aponeurosis

A. Assists flexion of wrist

N. Median nerve (C6, 7)

P. Tendon palpated in midline of anterior surface of wrist on radial side of flexor carpi ulnaris. When you flex your wrist against resistance or abduct your thumb, the palmaris longus, if present, will stand out in the middle of the flexor surface of your wrist.

Its long tendon is often used for tendon repair. Palmaris longus is sometimes absent on one or both sides. Palmaris brevis, a small muscle lying in the fascia of the hypothenar eminence, may be present.

Flexor Carpi Radialis

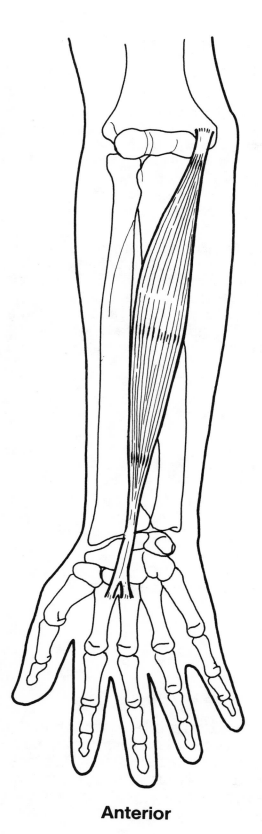

Anterior

O. Medial epicondyle of humerus

I. Bases of 2nd and 3rd metacarpals

A. Flexion
   Abduction of wrist

N. Median nerve (C6, 7)

P. Tendon palpated on anterior surface of wrist in line with 2nd metacarpal, just radial to palmaris longus tendon

Acting alone, this muscle flexes the wrist with some abduction. Flexor carpi radialis and extensor carpi radialis longus function together to abduct the wrist.

Flexor Digitorum Superficialis

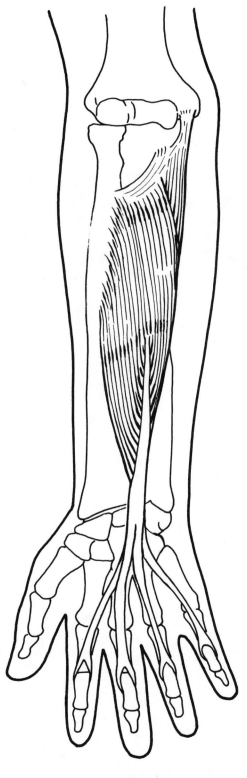

Anterior

O. Medial epicondyle of humerus
Coronoid process of ulna
Oblique line of radius

I. Sides of shafts of middle phalanges of 4 fingers

A. Flexion of 4 fingers at PIP joint

N. Median nerve (C7, 8, T1)

P. Tendon palpated on anterior surface of wrist on ulnar side between flexor carpi ulnaris tendon and palmaris longus tendon

The median nerve and ulnar artery pass beneath the origin of this muscle. At the wrist the tendons to the middle and ring fingers are anterior to the tendons to the index and little fingers. Each of the four digitorum tendons divides at the proximal phalanx to allow the flexor digitorum profundus to pass through to the distal phalanx.

Flexor Digitorum Profundus

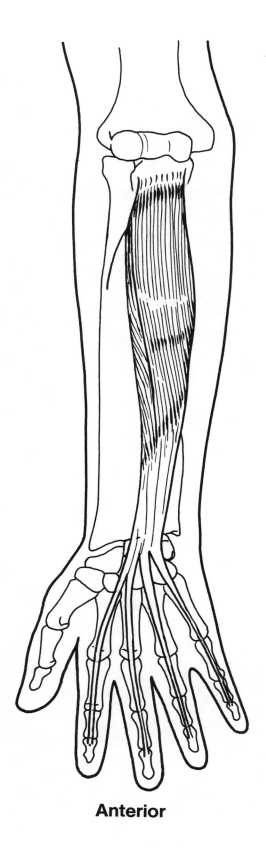

Anterior

O. Middle of anterior ulna and interosseous membrane

I. Bases of distal phalanges of 4 fingers

A. Flexion of 4 fingers at DIP joints

N. Median nerve to radial 2 fingers (C8, T1)
Ulnar nerve to ulnar 2 fingers (C8, T1)

P. Tendons palpated on anterior surfaces of middle phalanges of 4 fingers

Profundus means deep (Latin). The profundus tendons pass through the tendons of the flexor digitorum superficialis.

Flexor Pollicis Longus

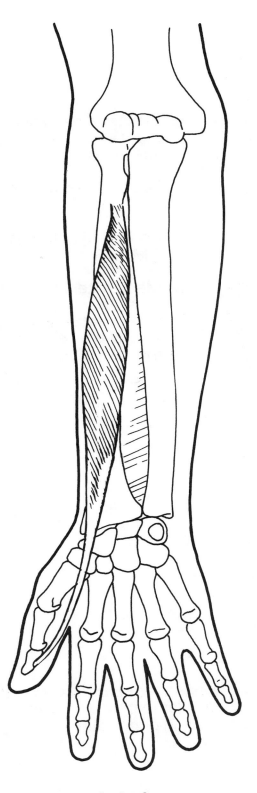

Anterior

O. Middle of anterior radius
   Interosseous membrane

I. Distal phalanx of thumb

A. Flexion of thumb at IP joint

N. Median nerve (C8, T1)

P. Tendon palpated on anterior surface of proximal phalanx

This muscle originates near the flexor digitorum profundus, and the muscle bellies are adjacent. The tendon passes through a synovial sheath and attaches to the distal phalanx of the thumb in a manner similar to the flexor digitorum profundus in the fingers.

Abductor Pollicis Brevis

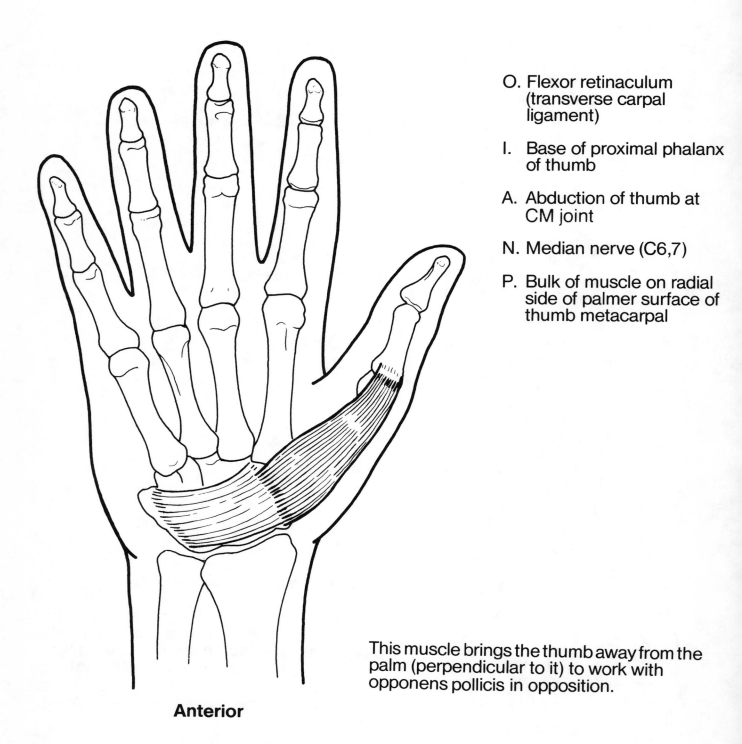

O. Flexor retinaculum (transverse carpal ligament)

I. Base of proximal phalanx of thumb

A. Abduction of thumb at CM joint

N. Median nerve (C6,7)

P. Bulk of muscle on radial side of palmer surface of thumb metacarpal

This muscle brings the thumb away from the palm (perpendicular to it) to work with opponens pollicis in opposition.

Anterior

Adductor Pollicis

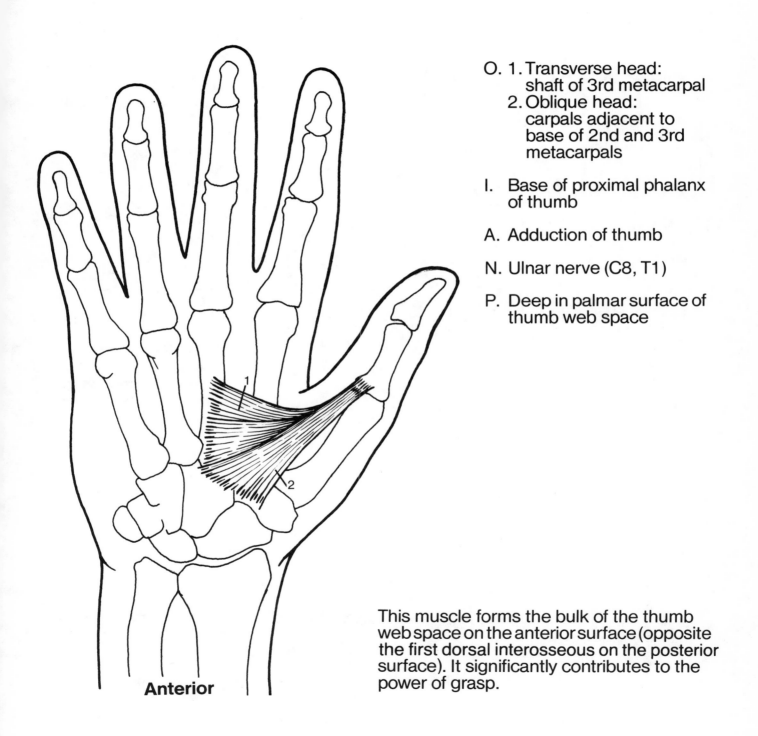

O. 1. Transverse head: shaft of 3rd metacarpal
2. Oblique head: carpals adjacent to base of 2nd and 3rd metacarpals

I. Base of proximal phalanx of thumb

A. Adduction of thumb

N. Ulnar nerve (C8, T1)

P. Deep in palmar surface of thumb web space

This muscle forms the bulk of the thumb web space on the anterior surface (opposite the first dorsal interosseous on the posterior surface). It significantly contributes to the power of grasp.

Flexor Pollicis Brevis

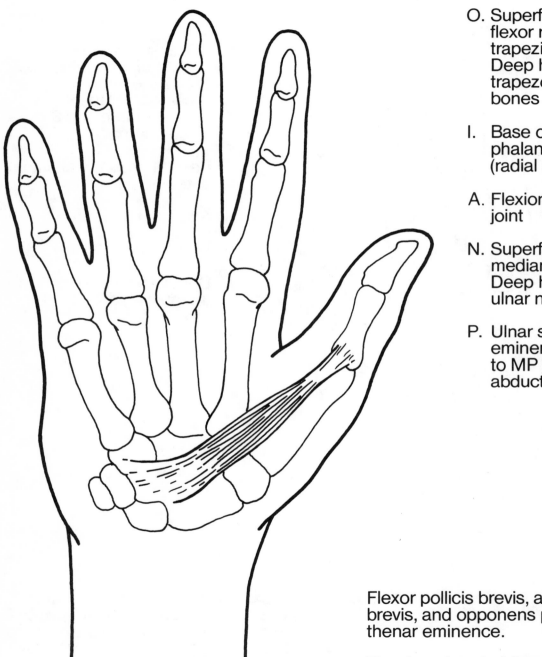

Anterior

O. Superficial head: flexor retinaculum and trapezium bone
   Deep head: trapezoid and capitate bones

I. Base of proximal phalanx of thumb (radial side)

A. Flexion of thumb at MP joint

N. Superficial head: median nerve (C6, 7)
   Deep head: ulnar nerve (C8, T1)

P. Ulnar side of thenar eminence just proximal to MP joint (medial to abductor policis brevis)

Flexor pollicis brevis, abductor pollicis brevis, and opponens pollicis form the thenar eminence.

The deep head of flexor pollicis brevis is sometimes considered as a palmer interosseous. (See palmer interossei.)

Flexor Digiti Minimi

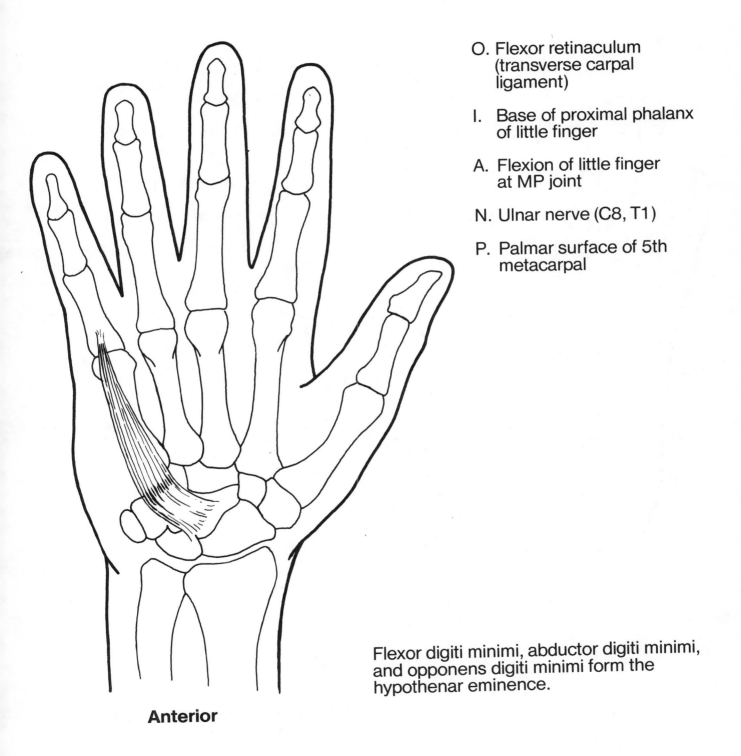

O. Flexor retinaculum (transverse carpal ligament)

I. Base of proximal phalanx of little finger

A. Flexion of little finger at MP joint

N. Ulnar nerve (C8, T1)

P. Palmar surface of 5th metacarpal

Flexor digiti minimi, abductor digiti minimi, and opponens digiti minimi form the hypothenar eminence.

Anterior

Opponens Pollicis

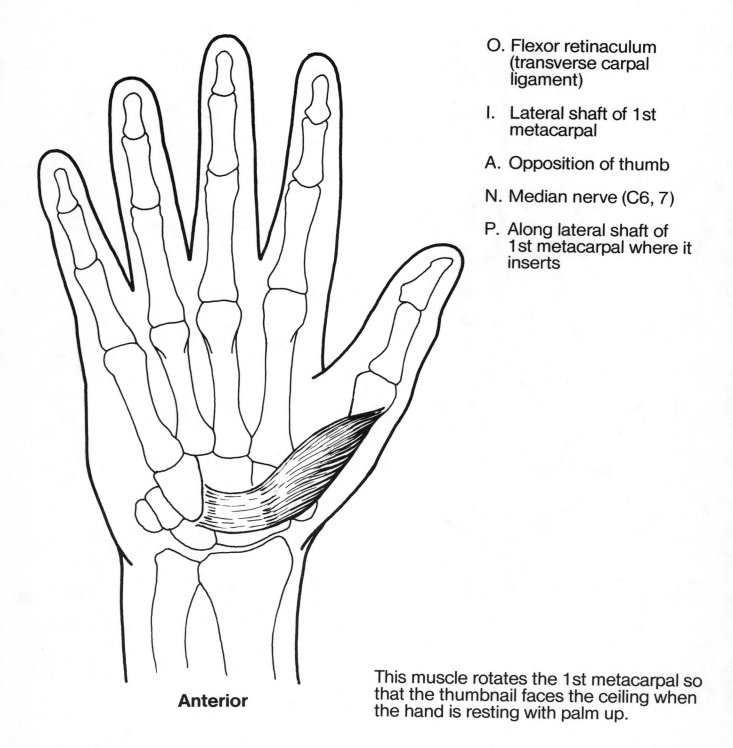

O. Flexor retinaculum (transverse carpal ligament)

I. Lateral shaft of 1st metacarpal

A. Opposition of thumb

N. Median nerve (C6, 7)

P. Along lateral shaft of 1st metacarpal where it inserts

Anterior

This muscle rotates the 1st metacarpal so that the thumbnail faces the ceiling when the hand is resting with palm up.

Opponens Digiti Minimi

O. Flexor retinaculum (transverse carpal ligament)

I. Ulnar border of 5th metacarpal

A. Opposition of little finger

N. Ulnar nerve (C8, T1)

P. Difficult to palpate

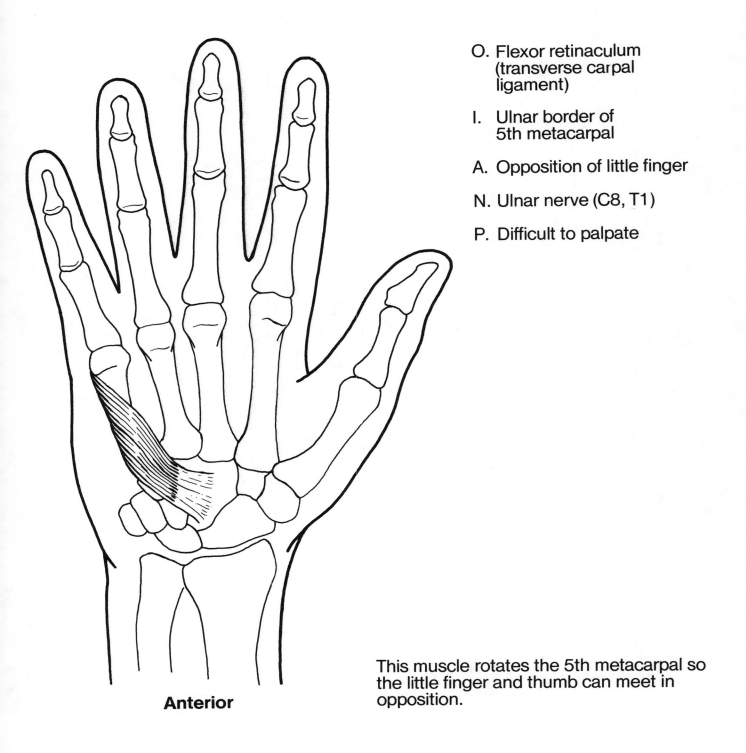

Anterior

This muscle rotates the 5th metacarpal so the little finger and thumb can meet in opposition.

Abductor Digiti Minimi

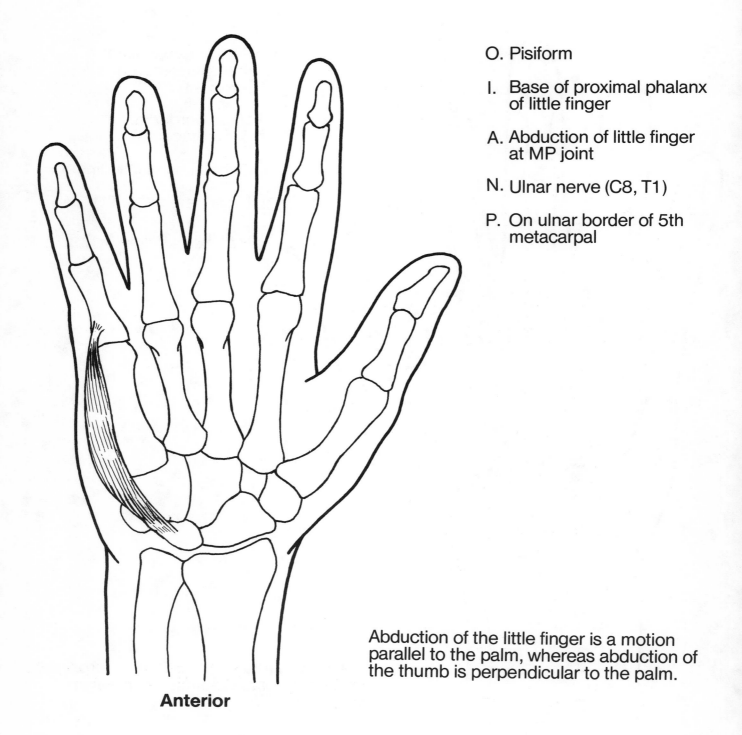

O. Pisiform

I. Base of proximal phalanx of little finger

A. Abduction of little finger at MP joint

N. Ulnar nerve (C8, T1)

P. On ulnar border of 5th metacarpal

Anterior

Abduction of the little finger is a motion parallel to the palm, whereas abduction of the thumb is perpendicular to the palm.

Dorsal Interossei (4)

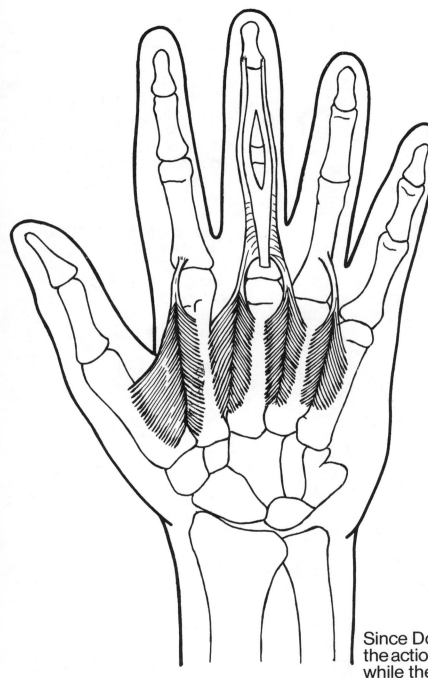

O. Metacarpals (adjacent surfaces)

I. Base of proximal phalanges to extensor expansion of 2nd, 3rd and 4th fingers

A. Abduction of 2nd, 3rd, and 4th fingers
Assists lumbricals in MP flexion and PIP and DIP extension

N. Ulnar nerve (C8, T1)

P. 1st: along radial side of 2nd metacarpal
Others: between metacarpals on dorsal surface

Posterior

Since Dorsal interossei ABduct the fingers, the action is recalled by the contraction DAB, while the Palmer interossei action of ADduction is differentiated as PAD.

Palmar Interossei (3)

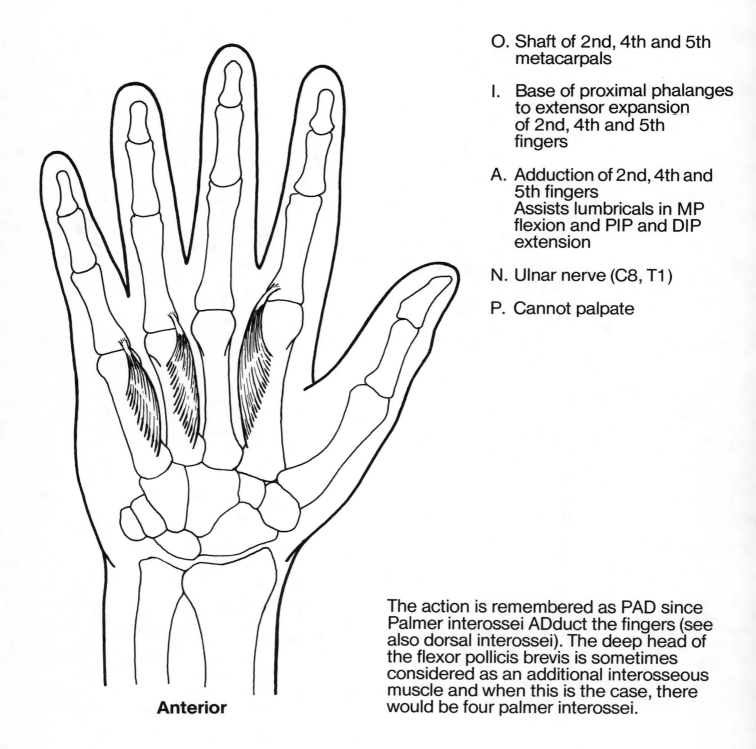

Anterior

O. Shaft of 2nd, 4th and 5th metacarpals

I. Base of proximal phalanges to extensor expansion of 2nd, 4th and 5th fingers

A. Adduction of 2nd, 4th and 5th fingers
Assists lumbricals in MP flexion and PIP and DIP extension

N. Ulnar nerve (C8, T1)

P. Cannot palpate

The action is remembered as PAD since Palmer interossei ADduct the fingers (see also dorsal interossei). The deep head of the flexor pollicis brevis is sometimes considered as an additional interosseous muscle and when this is the case, there would be four palmer interossei.

Lumbricals (4)

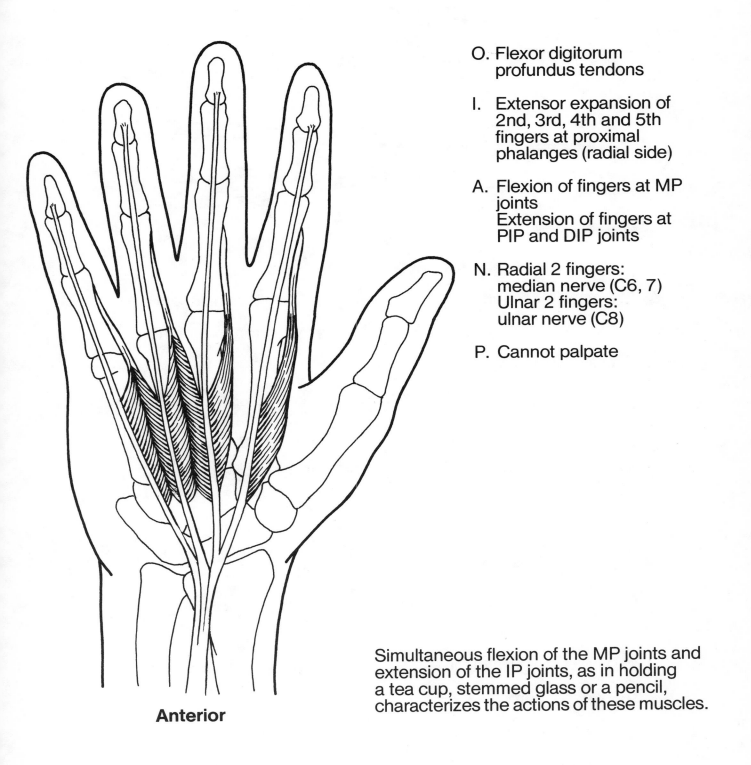

O. Flexor digitorum profundus tendons

I. Extensor expansion of 2nd, 3rd, 4th and 5th fingers at proximal phalanges (radial side)

A. Flexion of fingers at MP joints
Extension of fingers at PIP and DIP joints

N. Radial 2 fingers: median nerve (C6, 7)
Ulnar 2 fingers: ulnar nerve (C8)

P. Cannot palpate

Anterior

Simultaneous flexion of the MP joints and extension of the IP joints, as in holding a tea cup, stemmed glass or a pencil, characterizes the actions of these muscles.

Dorsal (Extensor) Expansion and Tendons Inserting on the Digits

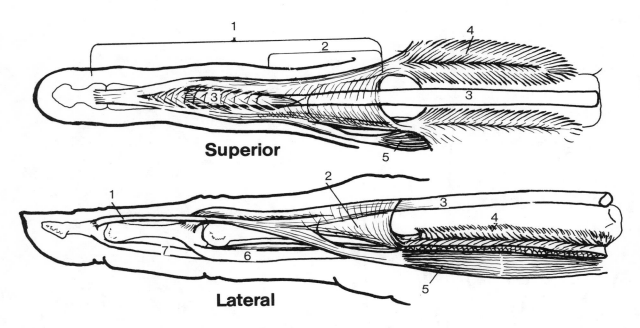

Superior

Lateral

1. Dorsal (extensor) expansion
2. Dorsal (extensor) hood
3. Extensor digitorum
4. Dorsal interosseous
5. Lumbrical
6. Flexor digitorum superficialis
7. Flexor digitorum profundus

Cross Section of Wrist

Dorsal

Ventral

1. Abductor pollicis longus
2. Extensor pollicis brevis
3. Extensor carpi radialis longus
4. Extensor carpi radialis brevis
5. Extensor pollicis longus
6. Extensor indicis
7. Extensor digitorum
8. Extensor digiti minimi
9. Extensor carpi ulnaris
10. Flexor carpi ulnaris
11. Flexor digitorum profundus
12. Flexor digitorum superficialis
13. Palmaris longus
14. Flexor pollicis longus
15. Flexor carpi radialis

Tendons of the Wrist and Digits

Posterior View

1. Abductor pollicis longus
2. Extensor pollicis brevis
3. Extensor pollicis longus
4. Extensor carpi radialis longus
5. Extensor carpi radialis brevis
6. Extensor digitorum
7. Extensor indicis
8. Extensor digiti minimi
9. Extensor carpi ulnaris

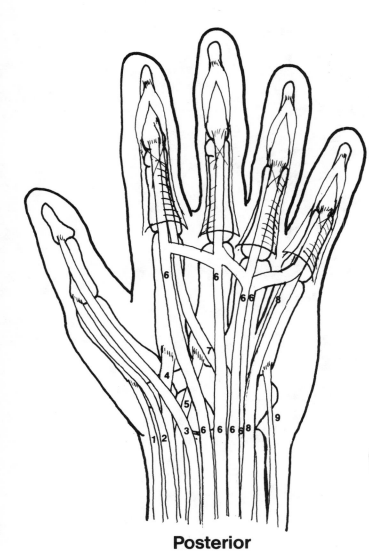

Posterior

Anterior View

1. Abductor pollicis longus
2. Flexor carpi radialis
3. Flexor pollicis longus
4. Palmaris longus tendon (cut away to reveal deeper structures)
5. Flexor digitorum superficialis
6. Flexor digitorum profundus
7. Flexor carpi ulnaris

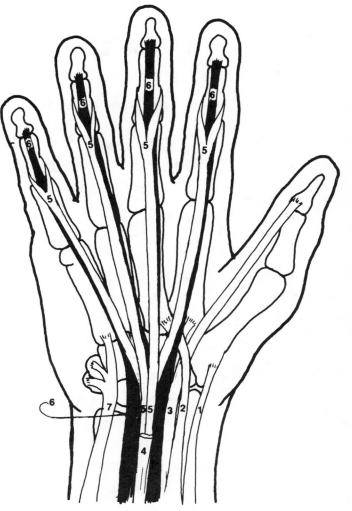

Anterior

57

Elevators of the Scapula

1. Upper trapezius
2. Levator scapula

Posterior

Depressors of the Scapula

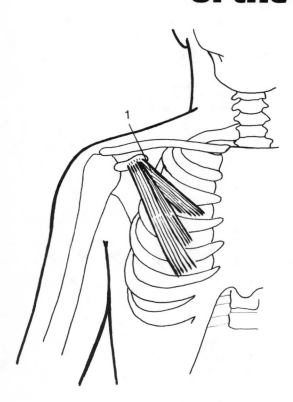

1. Pectoralis minor

Anterior

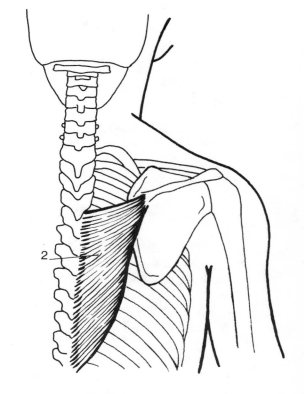

2. Lower trapezius

Posterior

Protractors of the Scapula

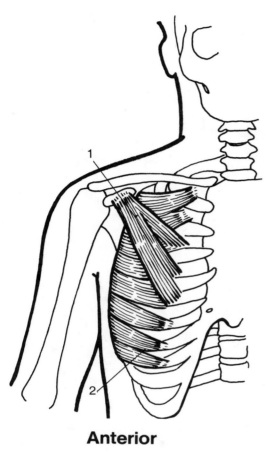

1. Pectoralis minor
2. Serratus anterior

Anterior

Retractors of the Scapula

1. Middle trapezius
2. Rhomboid

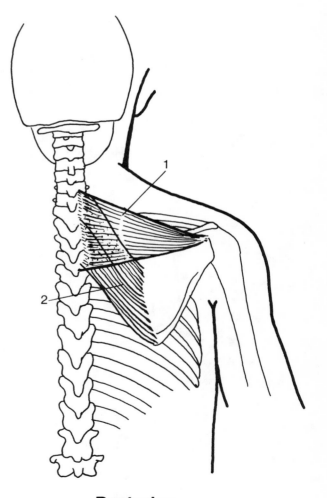

Posterior

Upward Rotators of the Scapula

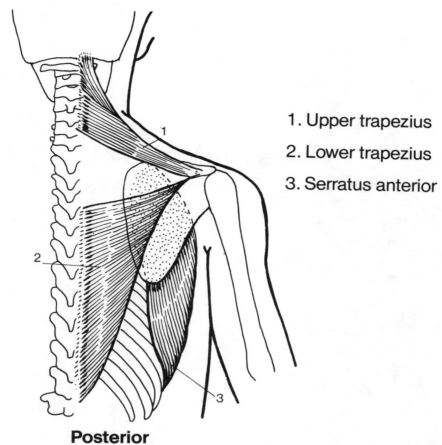

1. Upper trapezius
2. Lower trapezius
3. Serratus anterior

Posterior

Downward Rotators of the Scapula

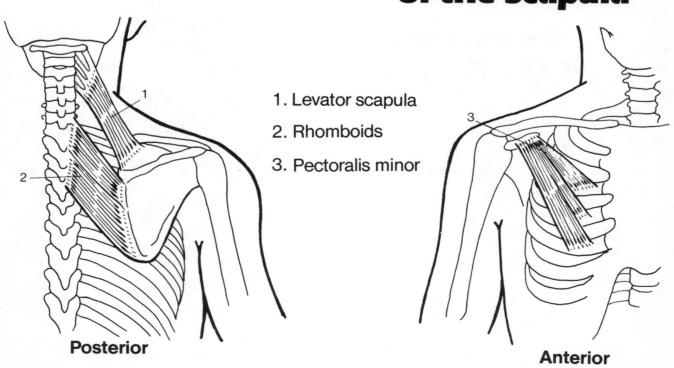

1. Levator scapula
2. Rhomboids
3. Pectoralis minor

Posterior **Anterior**

Medial (Internal) Rotators of the Humerus

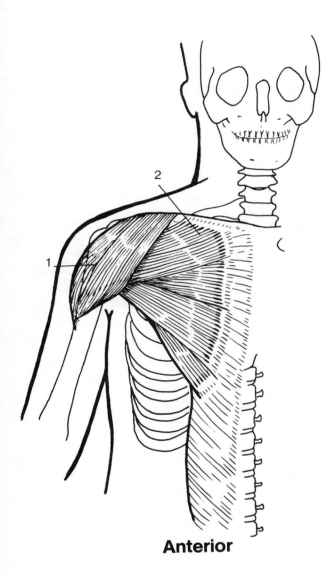

Anterior

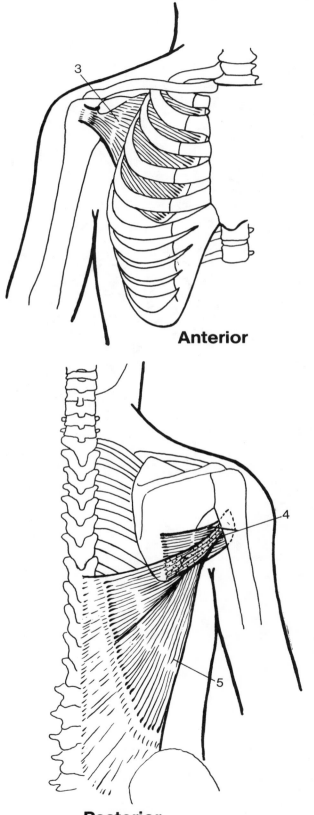

Anterior

Posterior

1. Anterior deltoid
2. Pectoralis major
3. Subscapularis
4. Teres major
5. Latissimus dorsi

Lateral (External) Rotation of the Humerus

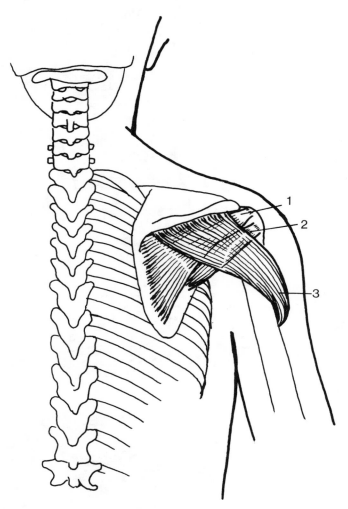

1. Infraspinatus
2. Teres minor
3. Posterior deltoid

Posterior

Flexors of the Humerus

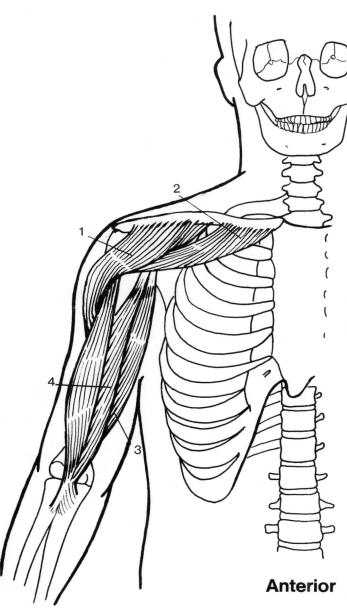

1. Anterior deltoid
2. Pectoralis major (clavicular head)
3. Coracobrachialis
4. Biceps (short head)

Anterior

Extensors of the Humerus

1. Latissimus dorsi
2. Teres major
3. Posterior deltoid
4. Infraspinatus
5. Teres minor
6. Triceps (long head)
7. Pectoralis major (sternal head)

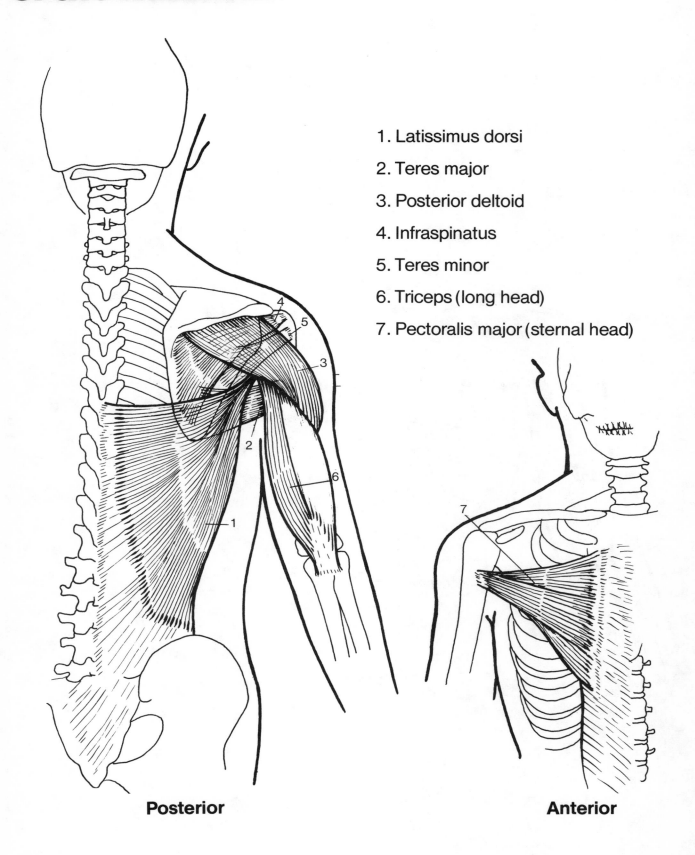

Posterior

Anterior

Abductors of the Humerus

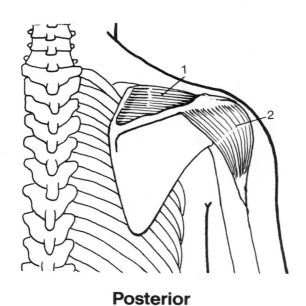

Posterior

1. Supraspinatus
2. Middle deltoid

Adductors of the Humerus

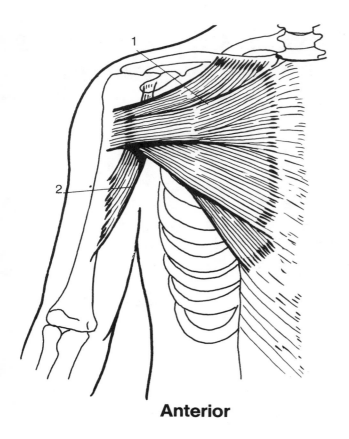

Anterior

1. Pectoralis major
2. Coracobrachialis

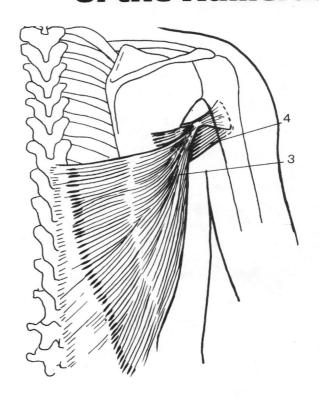

Posterior

3. Latissimus dorsi
4. Teres major

Horizontal Abductors of the Humerus

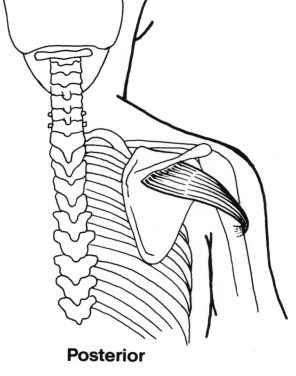

1. Posterior deltoid

Posterior

Horizontal Adductors of the Humerus

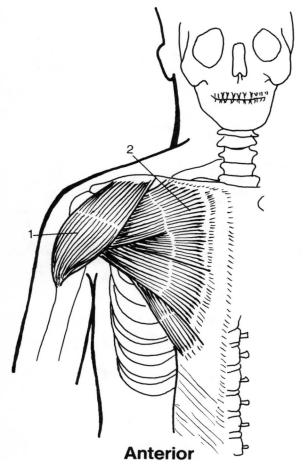

1. Anterior deltoid
2. Pectoralis major

Anterior

Flexors of the Elbow

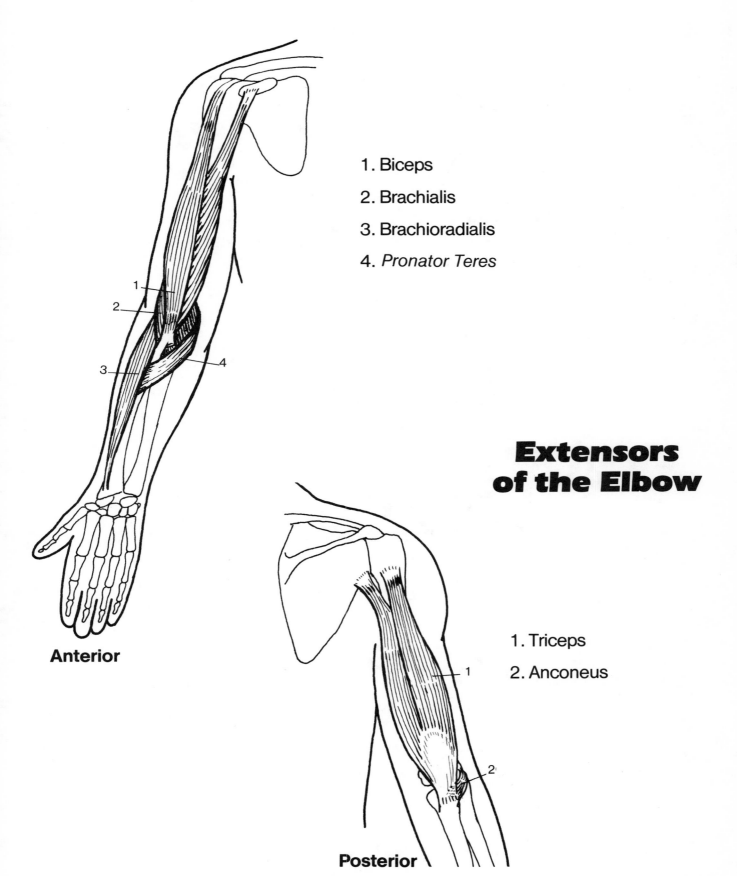

1. Biceps
2. Brachialis
3. Brachioradialis
4. *Pronator Teres*

Extensors of the Elbow

1. Triceps
2. Anconeus

Anterior

Posterior

Supinators of the Forearm

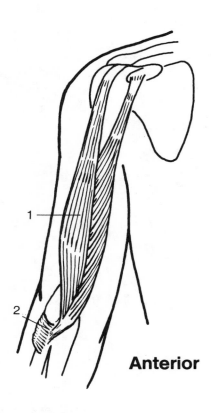

1. Biceps
2. Supinator

Anterior

Pronators of the Forearm

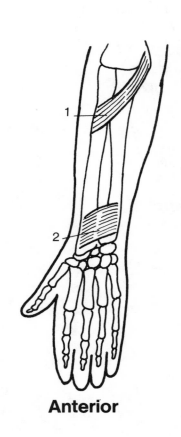

1. Pronator teres
2. Pronator quadratus

Anterior

Flexors of the Wrist

1. Flexor carpi radialis
2. Flexor carpi ulnaris
3. *Palmaris longus*

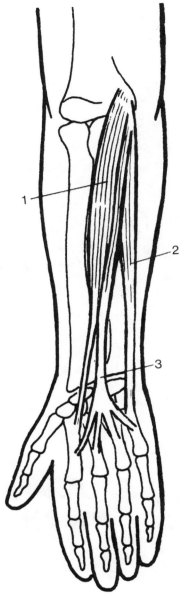

Anterior

Extensors of the Wrist

1. Extensor carpi radialis longus
2. Extensor carpi radialis brevis
3. Extensor carpi ulnaris

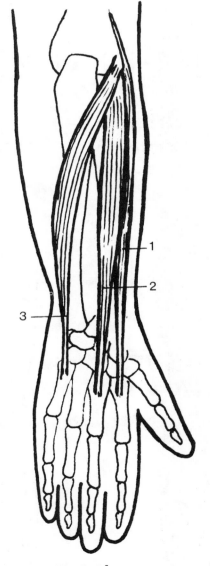

Posterior

Adductors of the Wrist

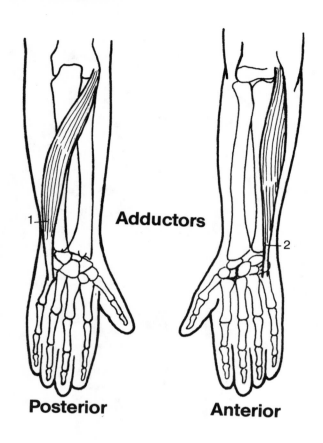

1. Extensor carpi ulnaris
2. Flexor carpi ulnaris

Abductors of the Wrist

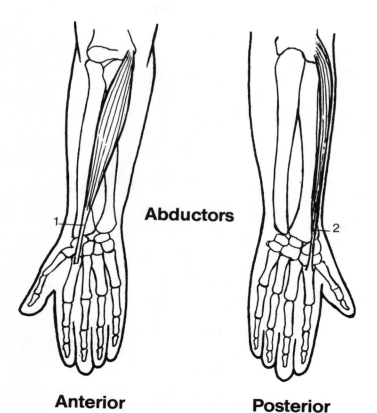

1. Flexor carpi radialis
2. Extensor carpi radialis longus

Abductors and Adductors of Thumb and Digits

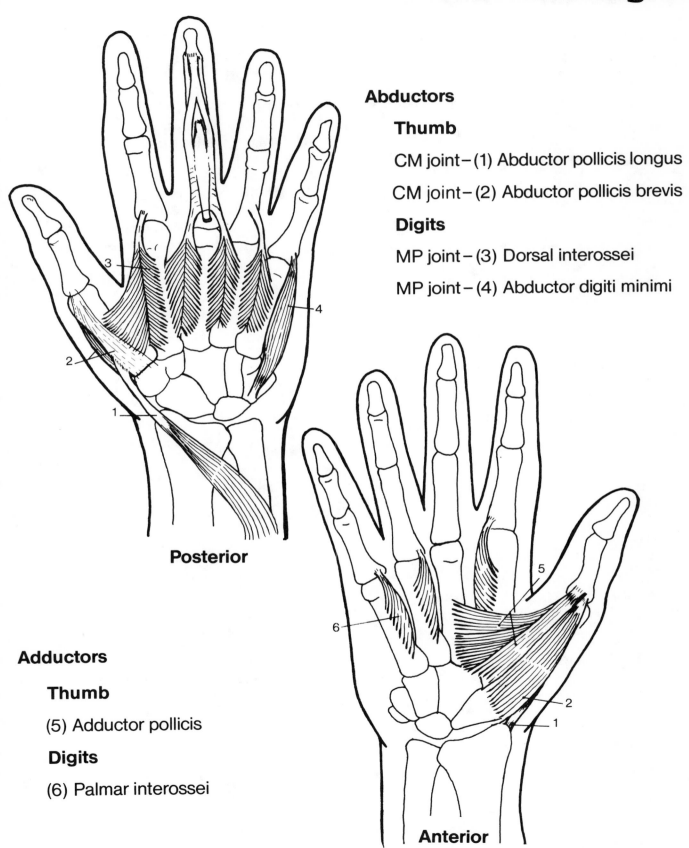

Abductors

  Thumb

  CM joint – (1) Abductor pollicis longus

  CM joint – (2) Abductor pollicis brevis

  Digits

  MP joint – (3) Dorsal interossei

  MP joint – (4) Abductor digiti minimi

Posterior

Adductors

  Thumb

  (5) Adductor pollicis

  Digits

  (6) Palmar interossei

Anterior

Extension, Flexion and Opposition of the Thumb and Digits

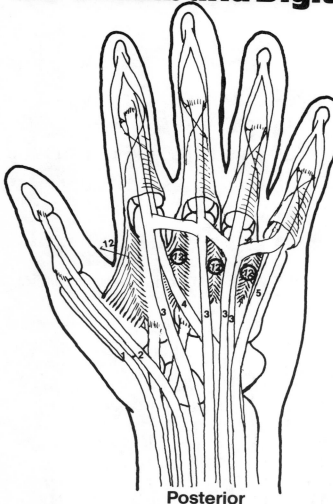

Posterior

Extension

Thumb
MP joint – extensor pollicis brevis (1)
IP joint – extensor pollicis longus (2)

Digits
MP joints – extensor digitorum (3)
     extensor indicis (4)
     extensor digiti minimi (5)
DIP & PIP joints – lumbricals (10)
     dorsal interossei (12)
     palmar interossei (11)

Flexion

Thumb
MP joint – flexor pollicis brevis (7)
IP joint – flexor pollicis longus (6)

Digits
MP joints – lumbricals (10)
     dorsal interossei (12)
     palmar interossei (11)
     flexor digiti minimi (not shown)
DIP joints – flexor digitorum profundus (8)
PIP joints – flexor digitorum
     superficialis (9)

Opposition

Thumb: opponens pollicis (13)

Little finger: opponens digiti minimi (14)

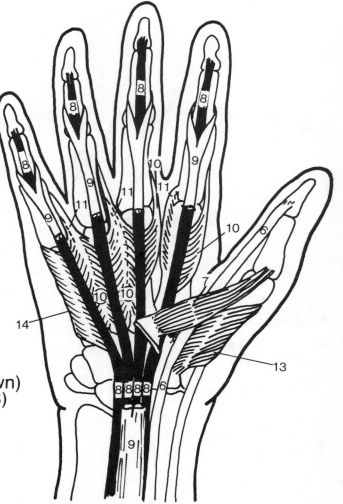

Anterior

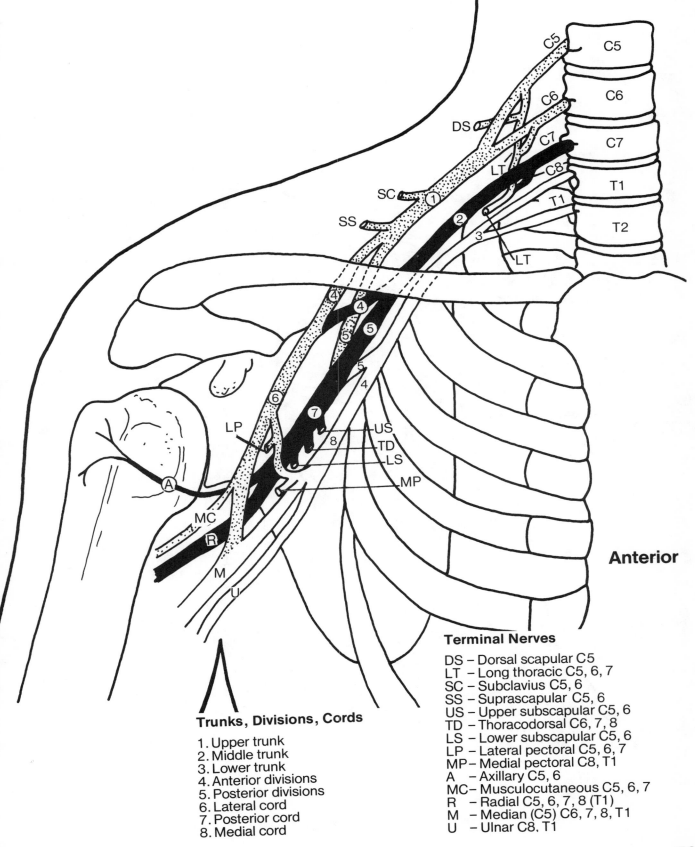

Axillary (Circumflex) Nerve

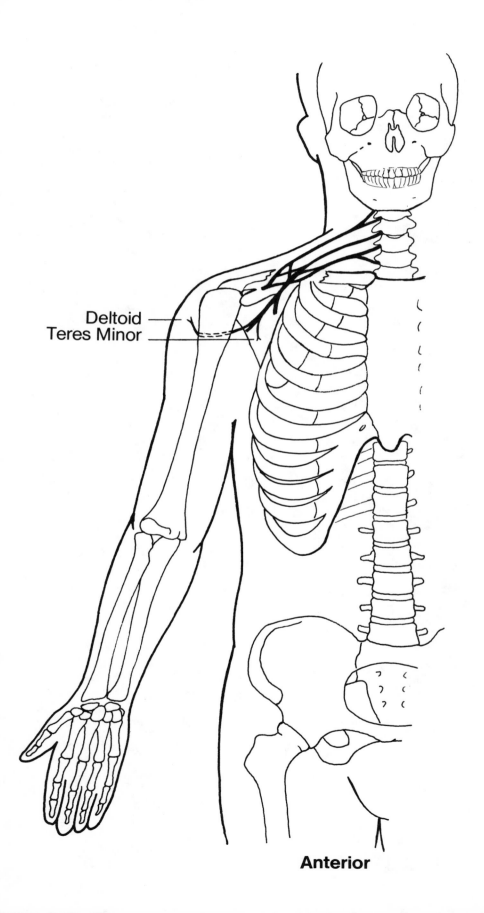

Anterior

Musculocutaneous Nerve

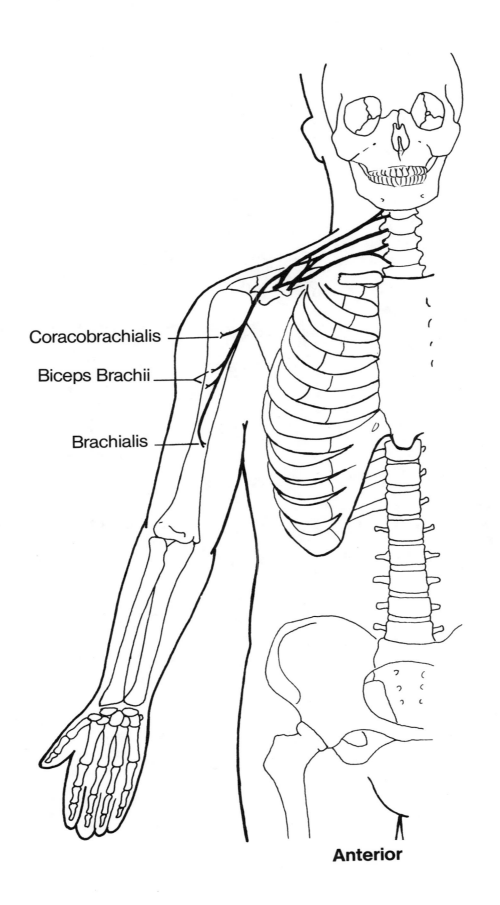

Anterior

Radial Nerve

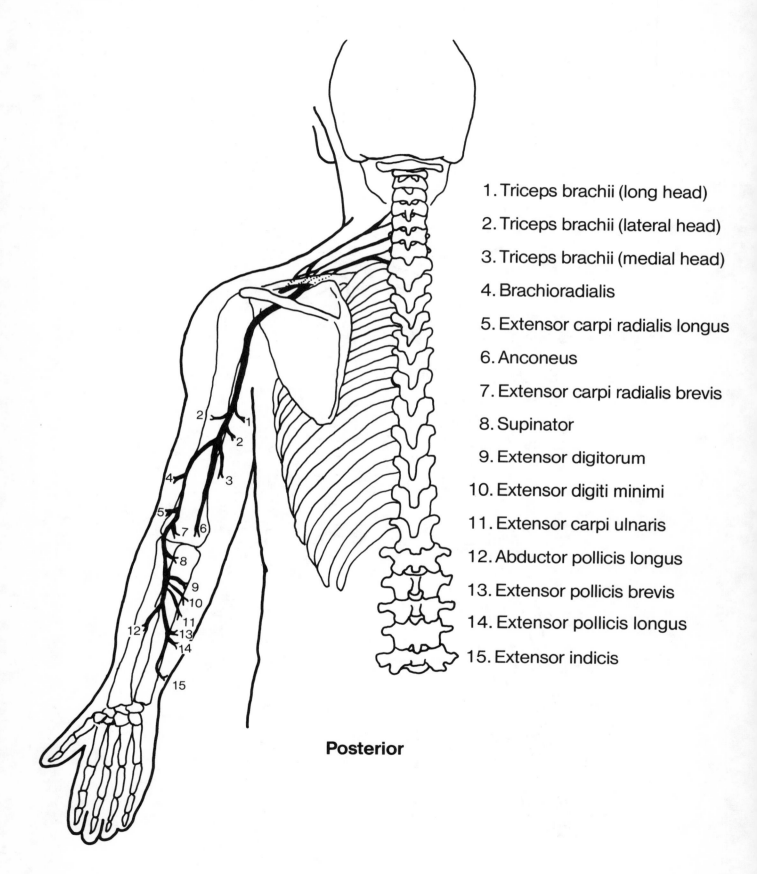

1. Triceps brachii (long head)
2. Triceps brachii (lateral head)
3. Triceps brachii (medial head)
4. Brachioradialis
5. Extensor carpi radialis longus
6. Anconeus
7. Extensor carpi radialis brevis
8. Supinator
9. Extensor digitorum
10. Extensor digiti minimi
11. Extensor carpi ulnaris
12. Abductor pollicis longus
13. Extensor pollicis brevis
14. Extensor pollicis longus
15. Extensor indicis

Posterior

Median Nerve

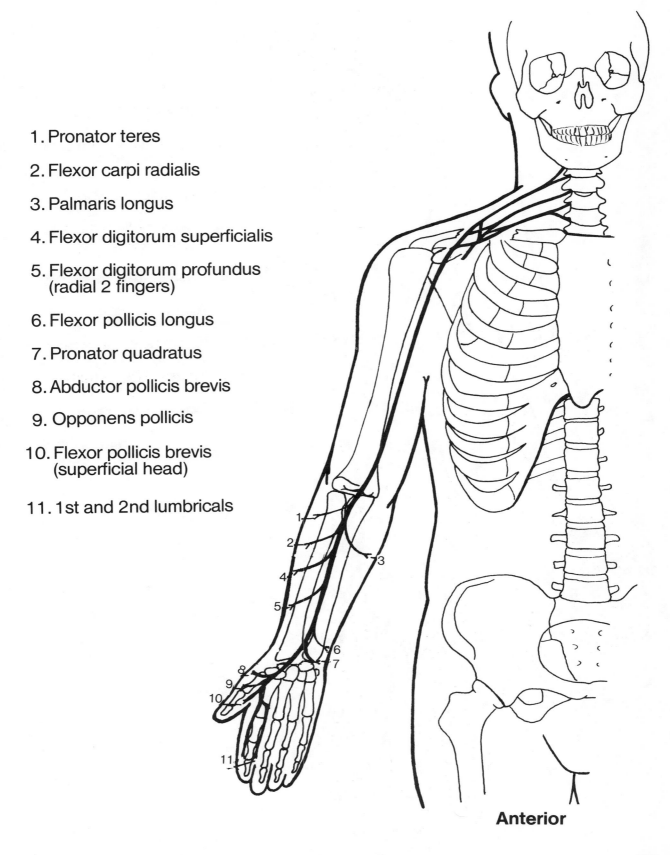

1. Pronator teres
2. Flexor carpi radialis
3. Palmaris longus
4. Flexor digitorum superficialis
5. Flexor digitorum profundus (radial 2 fingers)
6. Flexor pollicis longus
7. Pronator quadratus
8. Abductor pollicis brevis
9. Opponens pollicis
10. Flexor pollicis brevis (superficial head)
11. 1st and 2nd lumbricals

Anterior

Ulnar Nerve

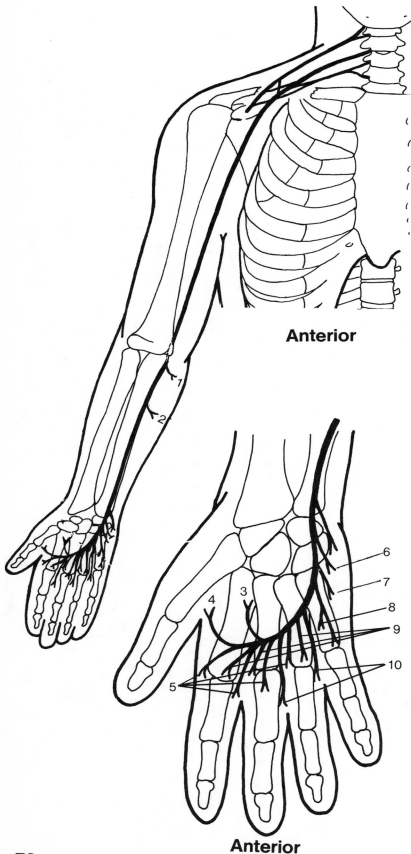

1. Flexor carpi ulnaris
2. Flexor digitorum profundus (medial 2)
3. Adductor pollicis
4. Flexor pollicis brevis (deep head)
5. Palmer interossei (3)
6. Abductor digiti minimi
7. Opponens digiti minimi
8. Flexor digiti minimi
9. Dorsal interossei (4)
10. 3rd and 4th lumbricals

Anterior

Anterior

Cutaneous Innervation of the Upper Limb

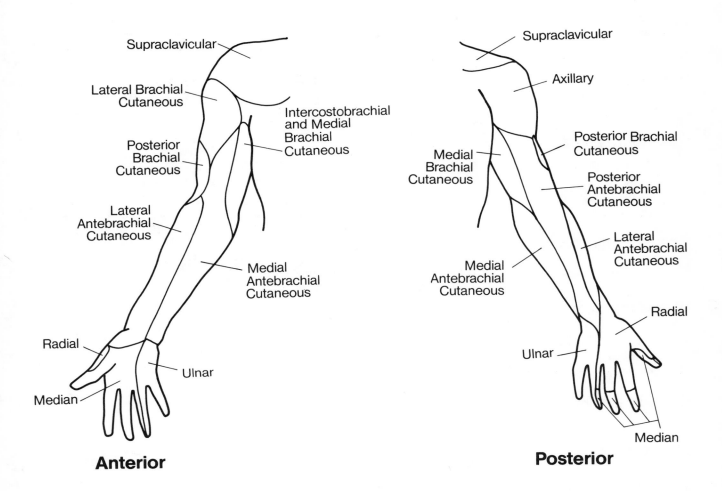

NERVE (Spinal Segments)	SOURCE
Supraclavicular (C3, 4)	Cervical plexus
Intercostobrachial (T2, 3)	Ventral rami T2, 3 Joins medial brachial cutaneous nerve
Medial brachial cutaneous (C8, T1)	Medial cord of brachial plexus
Axillary (C5, 6)	Terminal branches of axillary nerve
Lateral brachial cutaneous (C5, 6)	Axillary nerve
Posterior brachial cutaneous (T1, 2)	Radial nerve
Medial antebrachial cutaneous (C6, T1)	Medial cord of brachial plexus
Lateral antebrachial cutaneous (C5, 6)	Musculocutaneous nerve
Posterior antebrachial cutaneous (C5, 6, 7, 8)	Radial nerve
Ulnar (C8, T1)	Terminal branches of ulnar nerve
Median (C6, 7, 8)	Terminal branches of median nerve
Radial (C6, 7, 8)	Terminal branches of radial nerve

Pelvis

A. Ilium
1. Iliac crest
2. Superior (posterior) gluteal line
3. Middle (anterior) gluteal line
4. Inferior gluteal line
5. Anterior superior iliac spine
6. Anterior inferior iliac spine
7. Posterior superior iliac spine
8. Posterior inferior iliac spine
9. Greater sciatic notch

B. Ischium
10. Ischial tuberosity
11. Ischial spine
12. Lesser sciatic notch
13. Ramus of ischium

C. Pubis
14. Pubic symphysis
15. Body of pubis
16. Superior ramus of pubis
17. Inferior ramus of pubis

D. Acetabulum
E. Obturator foramen
F. Sacrum
G. Coccyx

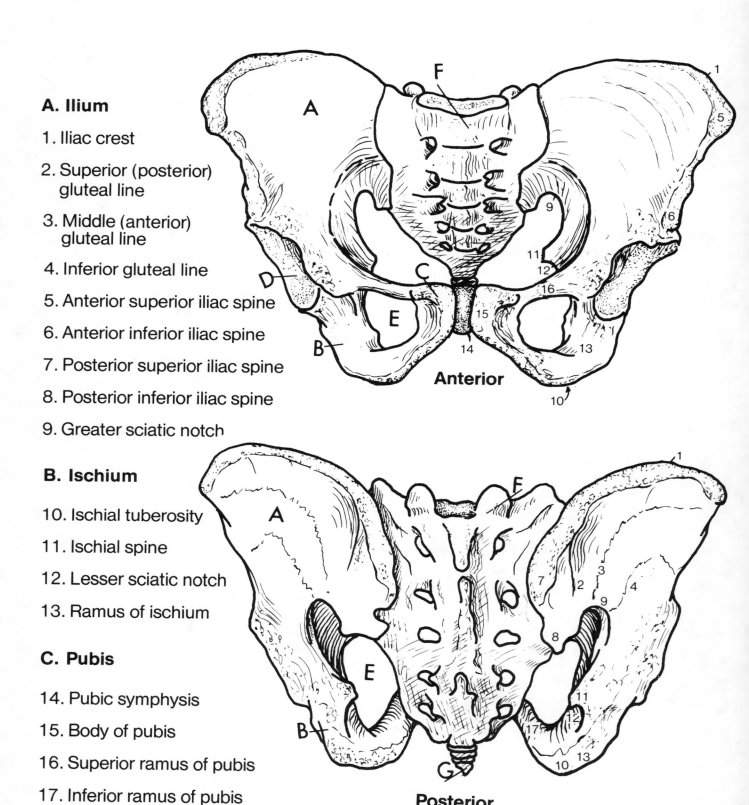

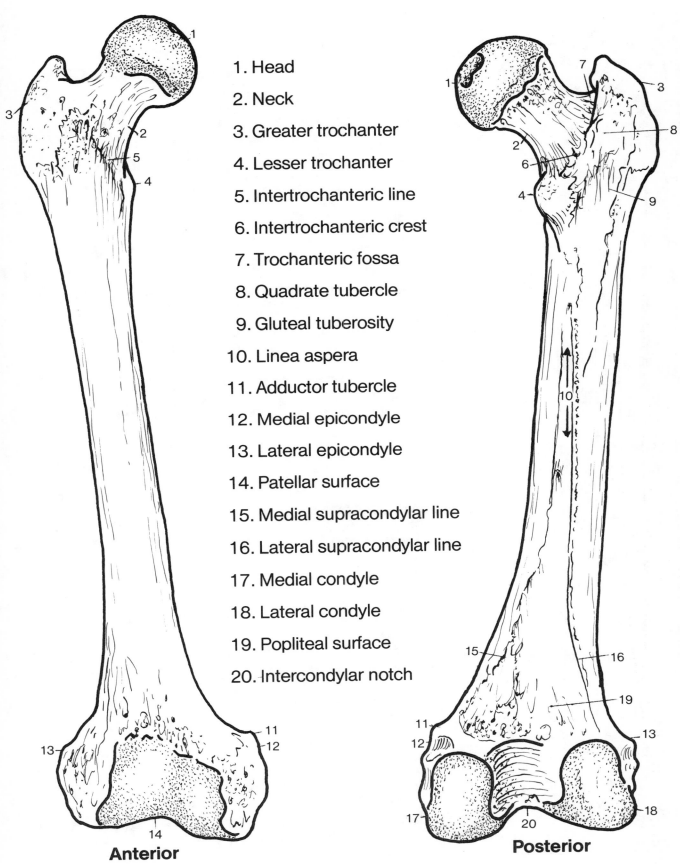

Right Femur

1. Head
2. Neck
3. Greater trochanter
4. Lesser trochanter
5. Intertrochanteric line
6. Intertrochanteric crest
7. Trochanteric fossa
8. Quadrate tubercle
9. Gluteal tuberosity
10. Linea aspera
11. Adductor tubercle
12. Medial epicondyle
13. Lateral epicondyle
14. Patellar surface
15. Medial supracondylar line
16. Lateral supracondylar line
17. Medial condyle
18. Lateral condyle
19. Popliteal surface
20. Intercondylar notch

Anterior

Posterior

Right Tibia and Fibula

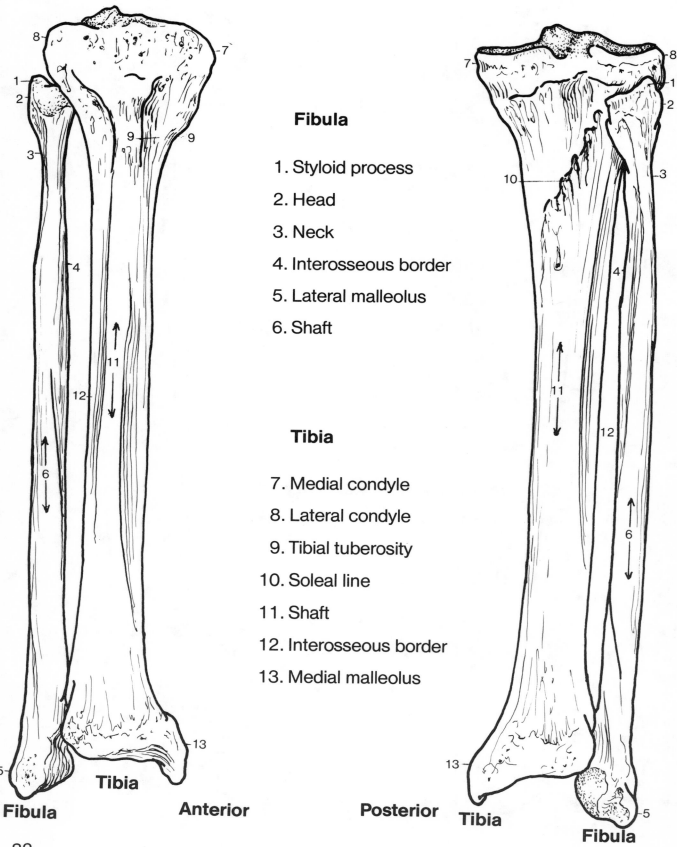

Fibula

1. Styloid process
2. Head
3. Neck
4. Interosseous border
5. Lateral malleolus
6. Shaft

Tibia

7. Medial condyle
8. Lateral condyle
9. Tibial tuberosity
10. Soleal line
11. Shaft
12. Interosseous border
13. Medial malleolus

Right Foot

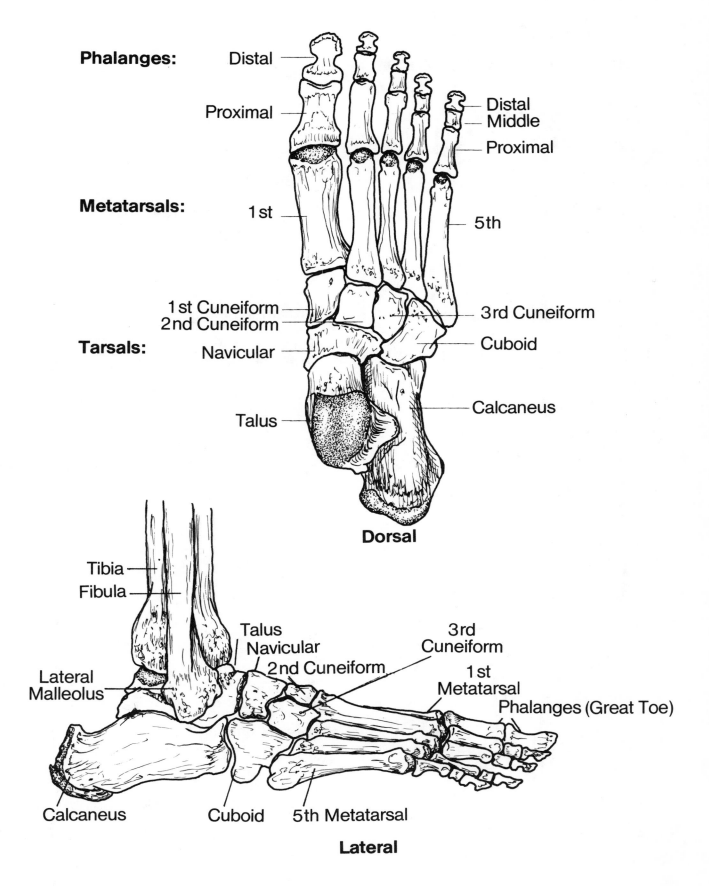

Palpation of Boney Landmarks
Pelvis and Lower Extremity

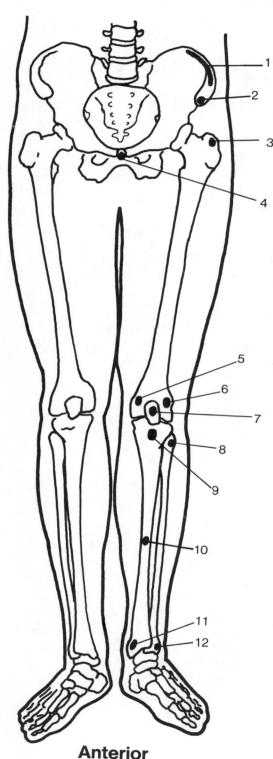

Anterior

1. Iliac crest – "hip bone" – From anterior superior iliac spine to posterior superior iliac spine

2. Anterior superior iliac spine – Anterior bump at end of iliac crest on anterior surface

3. Greater trochanter of femur – Large boney prominence on lateral proximal thigh at hip joint

4. Pubic symphysis – Anterior midline joint of pelvic girdle

5. Medial epicondyle of femur – Distal enlargement on medial femur

6. Lateral epicondyle of femur – Distal enlargement on lateral femur

7. Patella – "knee cap"

8. Head of fibula – Bump on superior end of fibula, inferior to lateral epicondyle of femur

9. Tuberosity of tibia – Large bump on proximal anterior tibia, inferior to patella

10. Anterior shaft of tibia – "shin bone" – Sharp margin below tibial tuberosity; can be palpated its entire length

11. Medial malleolus – Medial "ankle bone"; distal end of tibia

12. Lateral malleolus – Lateral "ankle bone"; distal pointed end of fibula

Palpation of Boney Landmarks

Pelvis and Lower Extremity

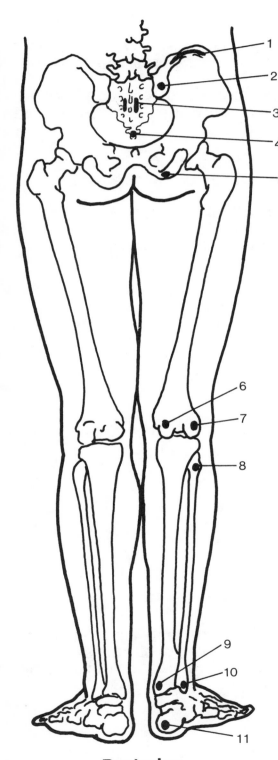

Posterior

1. Iliac crest – "hip bone" – From anterior superior iliac spine to posterior superior iliac spine

2. Posterior superior iliac spine – Boney prominence at end of iliac crest on posterior surface; lies just above the "dimple" or depression which is the landmark for the sacroiliac joint on the body surface

3. Sacrum – Curved triangular bone (five fused vertebrae) beneath lumbar spine

4. Coccyx – Caudal tip of vertebral column, deep between the two buttocks

5. Ischial tuberosity – Boney prominence that takes weight of body when seated upright

6. Medial epicondyle of femur – Distal enlargement on medial femur

7. Lateral epicondyle of femur – Distal enlargement on lateral femur

8. Head of fibula – Bump on superior end of fibula, inferior to lateral epicondyle of femur

9. Medial malleolus – Medial "ankle bone"; distal end of tibia

10. Lateral malleolus – Lateral "ankle bone"; distal pointed end of fibula

11. Calcaneus – "heel bone"

Gluteus Maximus

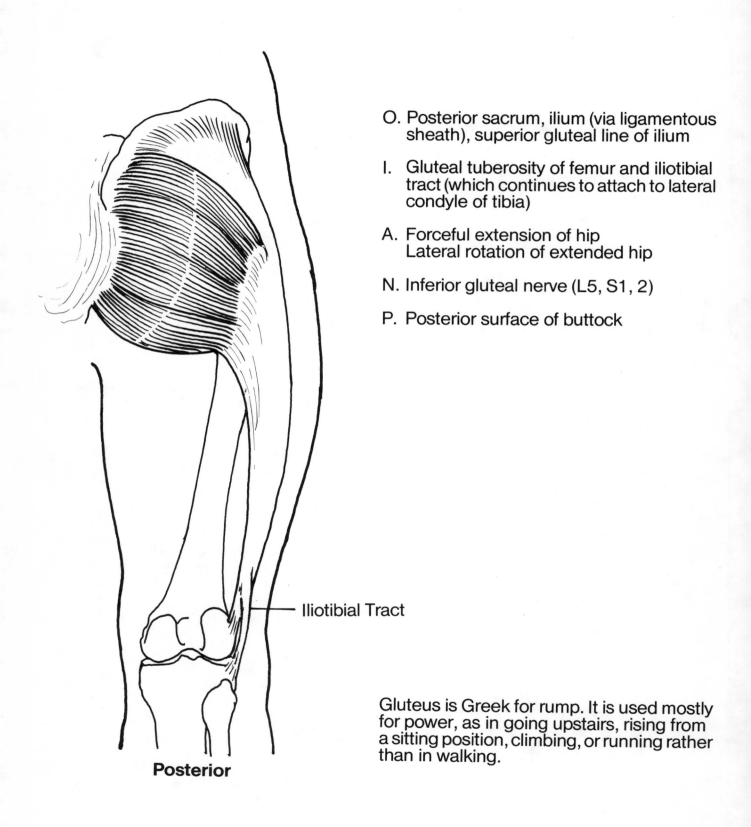

O. Posterior sacrum, ilium (via ligamentous sheath), superior gluteal line of ilium

I. Gluteal tuberosity of femur and iliotibial tract (which continues to attach to lateral condyle of tibia)

A. Forceful extension of hip
Lateral rotation of extended hip

N. Inferior gluteal nerve (L5, S1, 2)

P. Posterior surface of buttock

Iliotibial Tract

Posterior

Gluteus is Greek for rump. It is used mostly for power, as in going upstairs, rising from a sitting position, climbing, or running rather than in walking.

Gluteus Medius

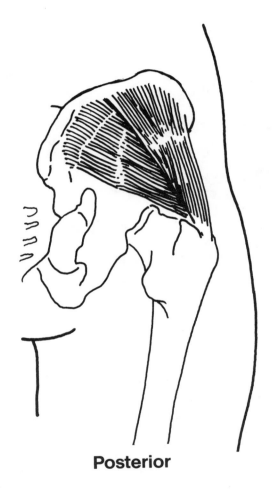

Posterior

O. Iliac crest; ilium between superior and middle gluteal lines

I. Greater trochanter of femur

A. Abduction
   Medial rotation of hip (anterior fibers)

N. Superior gluteal nerve (L4, 5, S1)

P. Lateral aspect of hip, above the greater trochanter

When standing on one foot, this muscle contracts on that side to keep the pelvis from tilting to the unsupported side. Alternate contraction of these muscles occurs in walking. Paralysis of this muscle on one side results in the "gluteus medius limp": the pelvis tilts towards the uninvolved side in walking.

Gluteus Minimus

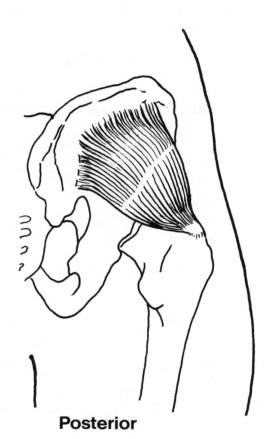

Posterior

O. Posterior ilium – between middle and inferior gluteal lines

I. Anterior surface of greater trochanter of femur

A. Abduction
   Medial rotation of hip

N. Superior gluteal nerve (L4, 5, S1)

P. With gluteus medius

Gluteus minimus works with gluteus medius.

Tensor Fasciae Latae

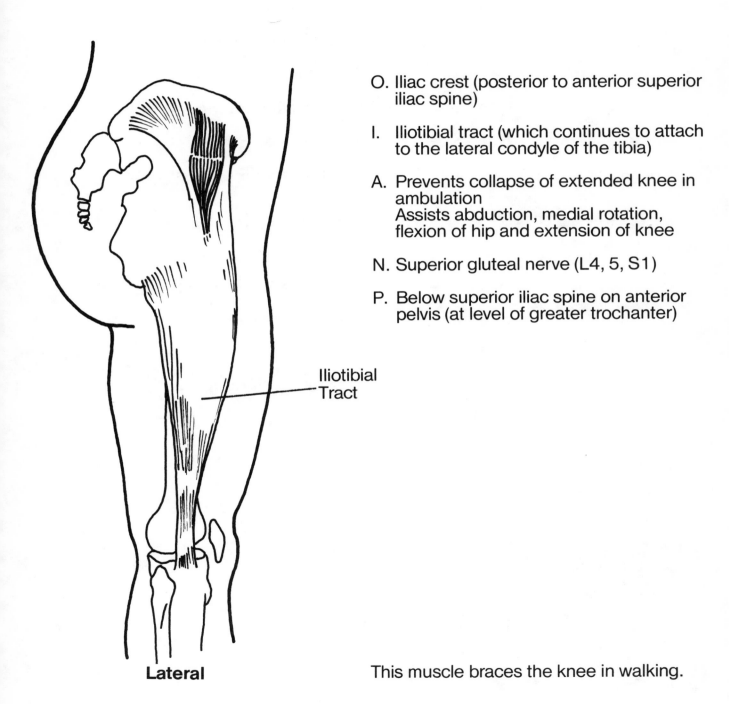

Lateral

Iliotibial Tract

O. Iliac crest (posterior to anterior superior iliac spine)

I. Iliotibial tract (which continues to attach to the lateral condyle of the tibia)

A. Prevents collapse of extended knee in ambulation
Assists abduction, medial rotation, flexion of hip and extension of knee

N. Superior gluteal nerve (L4, 5, S1)

P. Below superior iliac spine on anterior pelvis (at level of greater trochanter)

This muscle braces the knee in walking.

Six Deep Lateral Rotators of the Hip

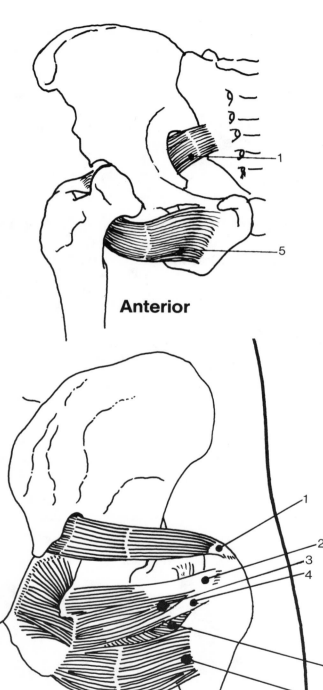

Anterior

Posterior

1. **Piriformis**
2. **Gemellus Superior**
3. **Obturator Internus**
4. **Gemellus Inferior**
5. **Obturator Externus**
6. **Quadratus Femoris**

O. Anterior sacrum
   Ischium
   Obturator foramen

I. Greater trochanter of femur

A. Lateral rotation of hip

N. Branches from sacral plexus (L4, 5, S1, 2)
   [Obturator externus supplied by obturator nerve (L3, 4)]

P. Cannot palpate

The first letter in the words of the phrase "Piece Goods Often Go On Quilts" denotes the anatomical order of these muscles from superior to inferior.

Psoas Major and Iliacus

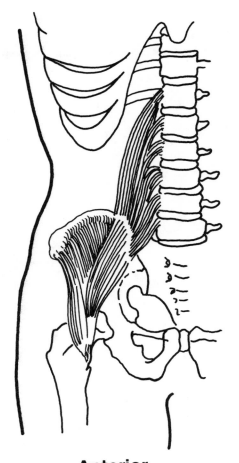

Anterior

O. Psoas major: lumbar vertebrae
   Iliacus: inner surface of ilium

I. Lesser trochanter of femur

A. Flexion, abduction and lateral rotation of hip

N. Iliacus – femoral nerve (L2, 3, 4)
   Psoas major – L2, 3 directly

P. Cannot palpate

Psoas major and iliacus are usually referred to as the iliopsoas muscle because of their common insertion and action. Iliopsoas is the strongest hip flexor. Psoas minor is not shown in drawing and is not present in most people. When present, it is a small muscle with a long tendon lying in front of psoas major, originating on the 12th thoracic vertebra, inserting on the pelvic brim and innervated by L1.

Sartorius

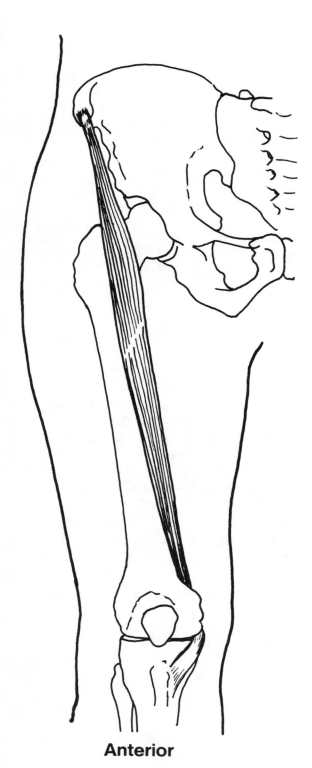

Anterior

O. Anterior superior iliac spine

I. Upper medial shaft of tibia

A. Assists flexion, abduction, lateral rotation of hip
Assists flexion, medial rotation of knee ("tailor position")

N. Femoral nerve (L2, 3, 4)

P. Close to its origin, just below anterior superior iliac spine, running diagonally across anterior pelvic area

The longest muscle in the body, it is the most superficial thigh muscle and forms the lateral border of the femoral triangle. Its name is derived from the Latin word for tailor, sartor, to indicate its action of bringing the leg into a cross-legged sitting position. Not a powerful muscle, it only assists in these actions.

Quadriceps Femoris Group

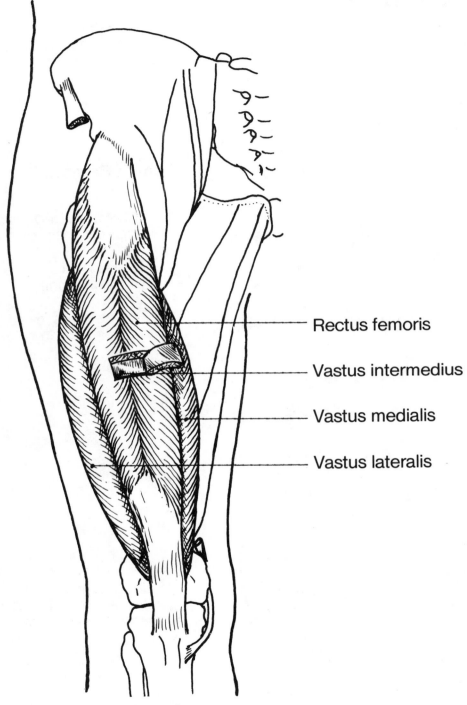

- Rectus femoris
- Vastus intermedius
- Vastus medialis
- Vastus lateralis

Anterior

These four, large, anterior thigh muscles insert below the knee and act to extend that joint. The three vasti lie deep to rectus femoris and two have their origins on the posterior femur. The rectus femoris originates on the pelvis and thereby can act to flex the hip. See the next pages for individual descriptions.

Quadriceps Group
(continued)

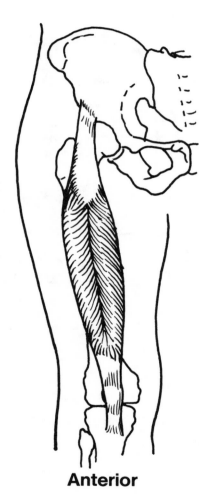

Anterior

Rectus Femoris

O. Anterior inferior iliac spine
   Upper margin of acetabulum

I. Patella and via patellar ligament to tibial tuberosity

A. Extension of knee
   Assists flexion of hip

N. Femoral nerve (L2, 3, 4)

P. Anterior surface of thigh

Rectus femoris is the only muscle in the quadriceps group that crosses both the hip and knee joints. Its combined actions are seen as the leg is brought forward in walking.

Quadriceps Group
(continued)

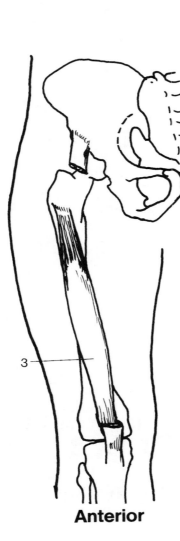

Anterior

Anterior

1. **Vastus Medialis**
2. **Vastus Lateralis**
3. **Vastus Intermedius**

O. Vastus medialis: linea aspera on posterior femur

Vastus lateralis: linea aspera on posterior femur

Vastus intermedius: anterior and lateral femoral shaft

I. Patella and via patellar ligament to tibial tuberosity

A. Extension of knee

N. Femoral nerve (L2, 3, 4)

P. Vastus medialis: anterior – medial surface of lower third of thigh, medial to biceps femoris

Vastus lateralis: lateral surface of thigh, lateral to biceps femoris

Vastus intermedius: cannot palpate

Vastus is the Latin term for immense. These three muscles of the quadriceps group derive their name from their size and position.

95

Pectineus

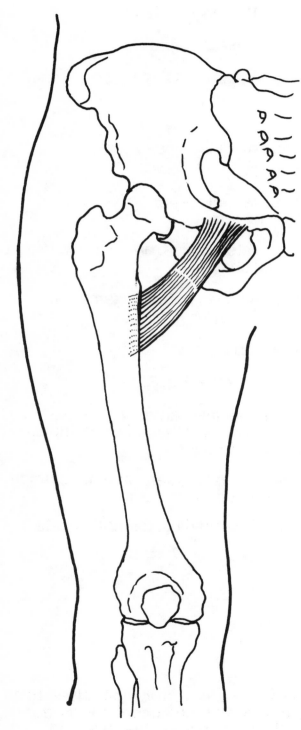

Anterior

O. Anterior pubis

I. Between lesser trochanter and linea aspera of posterior femur

A. Flexion of hip
Assists adduction and medial rotation of hip

N. Femoral nerve (L2, 3, 4)

P. Lateral and slightly superior to adductor longus on anterior pubis (difficult to differentiate from adductor longus)

This is the only adductor supplied by the femoral nerve, a fact explained by it being considered an extension of the iliopsoas. Pectineus is the uppermost muscle of the medial thigh muscles.

Adductor Longus and Adductor Brevis

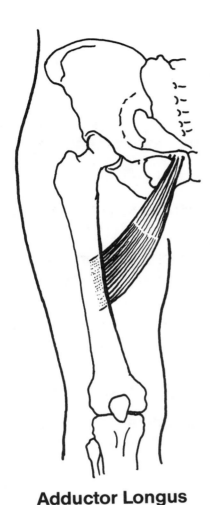

Adductor Longus   Anterior   **Adductor Brevis**

O. Anterior publis

I. Linea aspera on posterior femur

A. Adduction of hip
Assists flexion and medial rotation of the hip

N. Obturator (L3, 4)

P. Adductor longus: just below its proximal cord-like origin at medial aspect of groin

   Adductor brevis: cannot palpate

Adductor longus forms the medial border of the femoral triangle.

Adductor brevis is found deep to the adductor longus.

Adductor Magnus

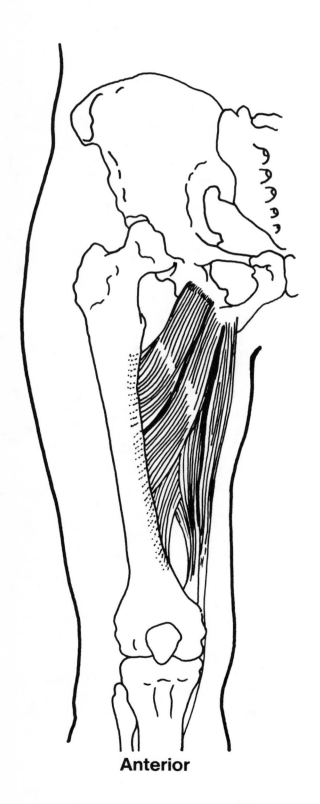

Anterior

O. Pubic ramus
   Ischial tuberosity

I. Linea aspera of posterior femur
   Adductor tubercle of medial femur

A. Adduction of hip

   Anterior fibers (which originate on pubic ramus) assist flexion of hip

   Posterior fibers (which originate on ischial tuberosity) assist extension of the hip

N. Anterior: obturator nerve (L2, 3, 4)
   Posterior: sciatic nerve (L4, 5, S1, 2, 3)

P. Medial surface of thigh

   Tendon palpated at its insertion on the adductor tubercle

This is the largest and deepest adductor.

Gracilis

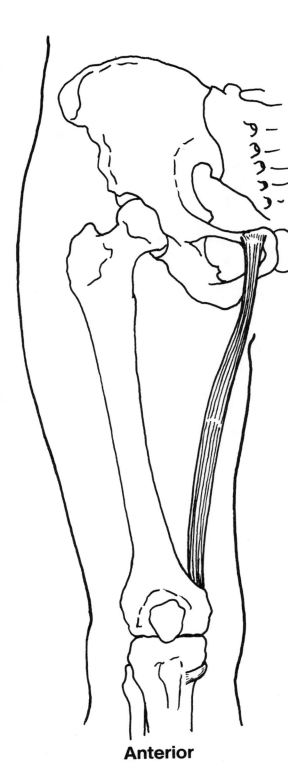

Anterior

O. Anterior pubis

I. Medial proximal tibia

A. Adduction of hip
Assists flexion and medial rotation of flexed knee

N. Obturator nerve (L2, 3, 4)

P. A few inches below public bone on medial side of thigh

Tendon palpated on medial side of posterior knee, medial to semitendinosus tendon

This is the most superficial and medial of the adductors. The femoral shaft and the gracilis form the shape of the letter "V".

Hamstrings

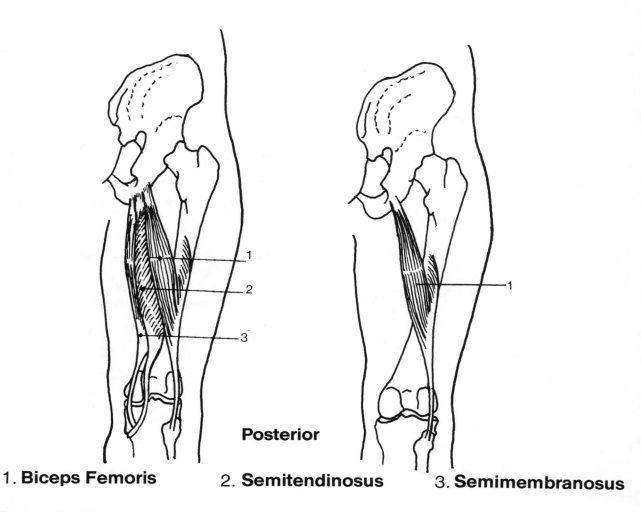

Posterior

1. **Biceps Femoris** 2. **Semitendinosus** 3. **Semimembranosus**

Biceps Femoris

O. Long head: ischial tuberosity
   Short head: linea aspera

I. Head of fibula

A. Long head: extension of hip
   Both heads: flexion of knee
                lateral rotation of flexed knee

N. Long head: Sciatic nerve – tibial division (S1, 2, 3)
   Short head: sciatic nerve – peroneal division (L5, S1, 2)

P. Lateral-posterior surface of thigh

   Tendon palpated on lateral aspect of posterior knee

All three of the hamstrings cross both the hip and knee joints. In order from lateral to medial their initials, BTM, indicate their arrangement on the "bottom" of the thigh. (See next page for individual descriptions of semitendinosus and semimembranosus.) Inability to touch the toes while keeping knees extended is largely due to shortened hamstrings.

Hamstrings
(continued)

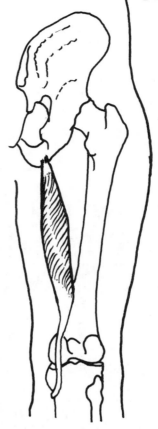

Posterior

Semimembranosus   **Semitendinosus**

O. Ischial tuberosity

I. Semimembranosus: posterior medial tibial condyle
Semitendinosus: anterior proximal tibial shaft

A. Extension of hip
Flexion of knee
Medial rotation of flexed knee

N. Sciatic nerve – tibial division (L5, S1, 2)

P. Semimembranosus: difficult to palpate because tendon is deep

Semitendinosus: tendon palpated on medial aspect of posterior knee (adjacent to gracilis tendon but lateral to it).

Semimembranosus and semitendinosus insert medially at the knee while the biceps femoris inserts laterally.

Gastrocnemius

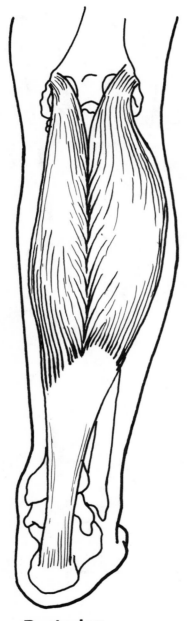

Posterior

O. Medial head: medial epicondyle of femur
   Lateral head: lateral epicondyle of femur

I. Calcaneus via tendo Achillis

A. Plantarflexion of ankle **or**
   Assists flexion of knee

N. Tibial nerve (S1, 2)

P. Upper half of posterior calf

   Tendon palpated as part of Achilles tendon

Gastro is the Greek term for "belly". This muscle can act on the knee or the ankle separately, but not simultaneously.

Soleus

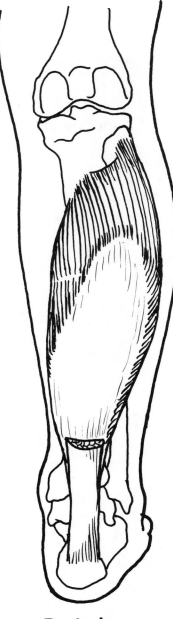

Posterior

O. Soleal line of tibia
   Posterior head and upper shaft of fibula

I. Calcaneus via tendo Achillis

A. Plantarflexion of ankle

N. Tibial nerve (S1, 2)

P. Lateral side of lower leg
   (below belly of gastrocnemius)

   Tendon palpated as part of Achilles tendon.

Soleus is Latin for sole, a flat fish. Deep to gastrocnemius, soleus is the stronger plantarflexor.

Plantaris

Posterior

O. Lateral epicondyle of femur

I. Calcaneus via tendo Achillis

A. Assists plantarflexion of ankle
Assists flexion of knee

N. Tibial nerve (L4, 5, S1)

P. Cannot palpate

Plantaris is the lower extremity counterpart of the palmaris longus and lies between the soleus and gastrocnemius. Gastrocnemius, soleus and plantaris share a common insertion, the tendo Achillis.

Popliteus

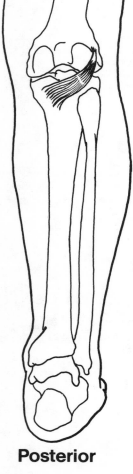

Posterior

O. Lateral condyle of femur

I. Posterior proximal tibial shaft

A. Initiates knee flexion by medial rotation of the tibia to "unlock" the extended knee

N. Tibial nerve (L5, S1)

P. Cannot palpate

Because of its action, popliteus is remembered as "the key that unlocks the knee". It is the deepest muscle at the back of the knee.

Tibialis Posterior

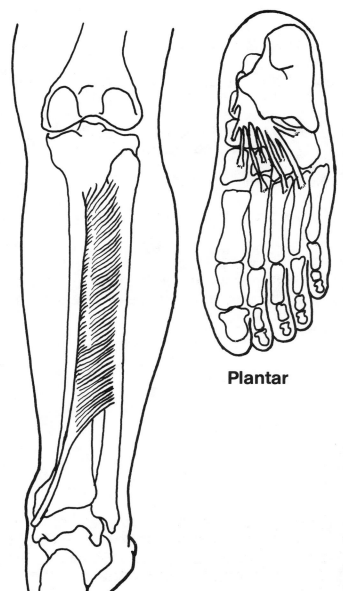

Posterior

Plantar

O. Posterior tibia, fibula, and interosseous membrane

I. Navicular bone, adjacent tarsals and metatarsals on plantar surface

A. Inversion of foot
   Assists plantarflexion of ankle

N. Tibial nerve (L5, S1)

P. Tendon palpated on medial malleolus

Three deep posterior calf muscles have tendons which course around medial malleolus, with the tendon of tibialis posterior being the most anterior and superficial, flexor digitorum next and flexor hallucis longus the most posterior and deep in relationship to medial malleolus. This order of the tendons is remembered by the first letters in the names "Tom Dick, and Harry" with the tibialis posterior being "Tom." Tibialis posterior is the deepest of the three muscles.

Flexor Digitorum Longus

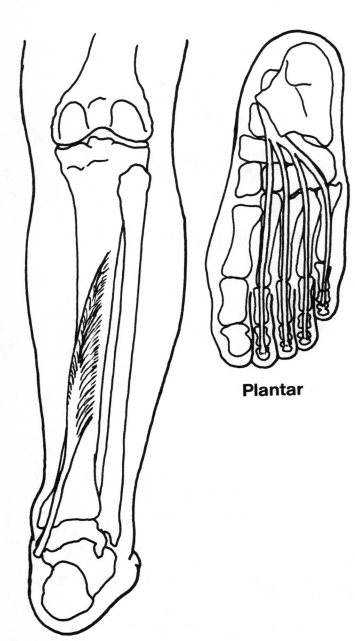

Posterior

Plantar

O. Posterior tibia

I. Distal phalanges of 4 lateral toes on plantar surface

A. Flexion of 4 lateral toes at DIP joints
Assists plantar flexion of ankle

N. Tibial nerve (L5, S1)

P. Medial aspect of distal calf

Tendon palpated going around medial malleolus just posterior to tibialis posterior tendon. (Alternate inversion and toe flexion to differentiate them.)

This muscle is recalled as "Dick" of Tom, Dick and Harry calf muscles. (See tibialis posterior and flexor hallucis longus.) Flexor digitorum longus is comparable to the flexor digitorum profundus in the hand.

Flexor Hallucis Longus

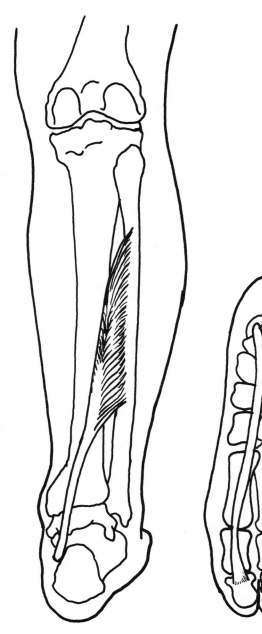

O. Posterior fibula

I. Distal phalanx of great toe (plantar surface)

A. Flexion of great toe
Assists plantarflexion of ankle

N. Tibial nerve (L5, S1, 2)

P. Tendon is difficult to differentiate from flexor digitorum longus. However, tendon may be palpated just medial and slightly deep to the tendo Achillis.

Flexor hallucis longus is "Harry" of Tom, Dick and Harry calf muscles. (See tibialis posterior and flexor digitorum longus.)

Tibialis Anterior

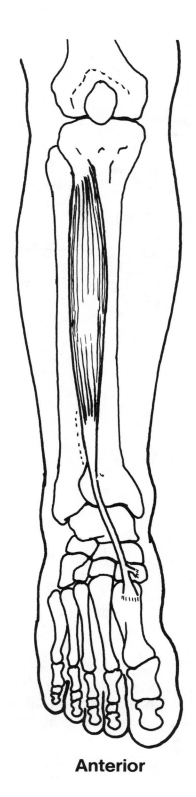

Anterior

O. Lateral shaft of tibia
   Interosseous membrane

I. Base of 1st metatarsal
   First (medial) cuneiform

A. Dorsiflexion of ankle
   Inversion of foot

N. Deep peroneal nerve (L4, 5, S1)

P. Lateral side of tibia on anterior surface

   Tendon palpated on medial side of anterior surface of ankle

Paralysis of this muscle causes foot drop.

Extensor Hallucis Longus

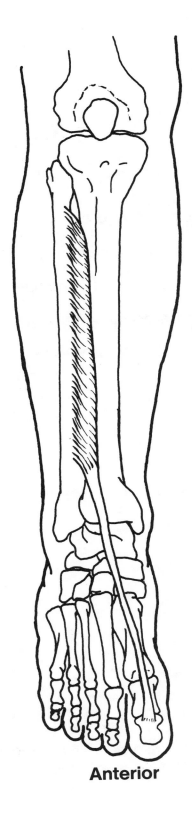

Anterior

O. Anterior shaft of fibula
   Interosseous membrane

I. Base of distal phalanx of the great toe

A. Extension of great toe
   Assists dorsiflexion of ankle

N. Deep peroneal nerve (L4, 5, S1)

P. Tendon palpated lateral to tibialis anterior tendon on anterior surface of ankle and also on dorsum of foot near the great toe

This muscle is comparable to the extensor pollicis longus in the hand.

Extensor Digitorum Brevis

Anterior

O. Anterior calcaneus

I. Extensor expansion of 4 medial toes

A. Aids extension of 4 medial toes

N. Deep peroneal nerve (L4, 5, S1)

P. Anterior to and slightly below lateral malleolus on dorsum of foot

The portion to the great toe is sometimes identified as the extensor hallucis brevis.

Extensor Digitorum Longus

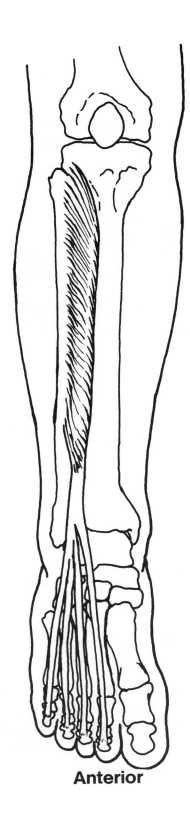

Anterior

O. Lateral condyle of tibia
Proximal ⅔ of anterior shaft of fibula

I. Middle and distal phalanges of 4 lateral toes

A. Extension of 4 lateral toes
Assists dorsiflexion of ankle

N. Deep peroneal nerve (L4, 5, S1)

P. Common tendon palpated on anterior surface of ankle, lateral to extensor hallucis longus tendon. The divided tendons palpated on the dorsum of the foot.

This toe extensor is comparable to the extensor digitorum communis in the hand.

Peroneus Tertius

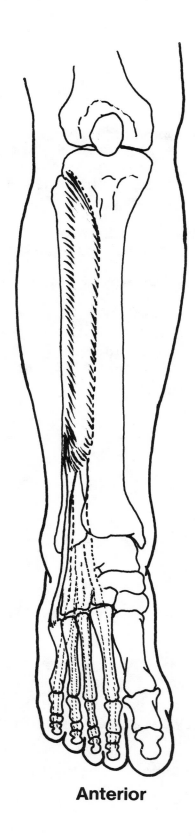

Anterior

O. Anterior distal fibula
(with extensor digitorum longus)

I. Base of 5th metatarsal

A. Eversion of foot
Assists dorsiflexion

N. Deep peroneal nerve (L4, 5, S1)

P. Tendon palpated lateral to extensor digitorum longus tendon on dorsum of foot at base of 5th metatarsal

This muscle functions to place the foot flat on the ground by raising the lateral border.

Peroneus Brevis

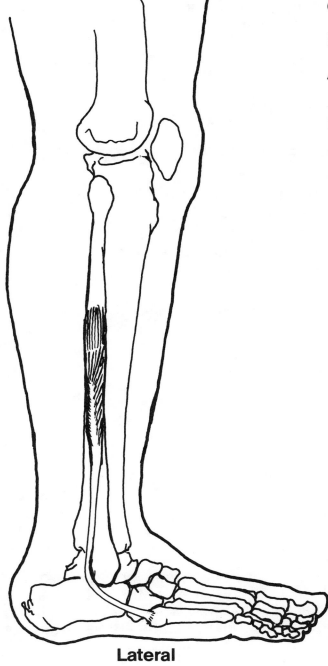
Lateral

O. Lateral shaft of fibula (lower ⅔)

I. Base of 5th metatarsal

A. Eversion of foot
Assists plantar flexion of ankle

N. Superficial peroneal nerve (L4, 5, S1)

P. Tendon palpated on lateral dorsum of foot where it inserts on the tuberosity at proximal end of 5th metatarsal. It is closest to the malleolus and stands out more than peroneus longus tendon.

The action of the foot evertors (as well as invertors) is especially helpful when walking or running on uneven surfaces. Sometimes the muscles "give out," and a sprained ligament can result.

Peroneus Longus

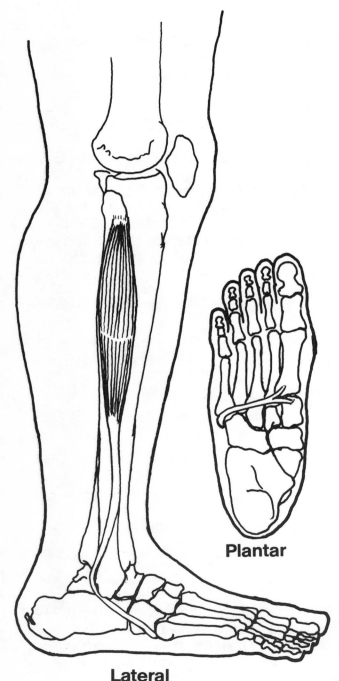

O. Lateral shaft of fibula (upper ⅔)

I. Base of 1st metatarsal and 1st (medial) cuneiform (plantar surface)

A. Eversion of foot
Assists plantar flexion of ankle

N. Superficial peroneal nerve (L4, 5, S1)

P. Lateral surface of proximal half of lower leg

Tendon palpated just above and behind lateral malleolus, slightly posterior to peroneus brevis tendon

Peroneus is derived from the Greek word for fibula, indicating the location of the peronei muscles. Peroneus longus traverses the sole of the foot to meet the tibialis anterior tendon to form a stirrup for the foot.

Muscles of the Foot
Layer 1

1. **Abductor Hallucis**
2. **Flexor Digitorum Brevis**
3. **Abductor Digiti Minimi**

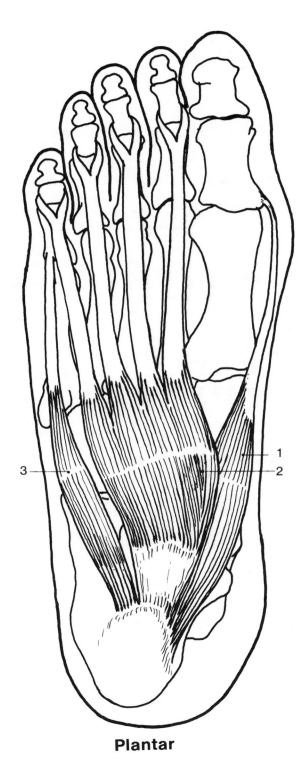

Plantar

1. **Abductor Hallucis**

O. Calcaneus

I. Base of proximal phalanx of great toe

A. Flexion, abduction of great toe at MP joint

N. Medial plantar nerve (L4, 5)

P. Cannot palpate

Comparable to the abductor pollicis brevis in the hand.

2. **Flexor Digitorum Brevis**

O. Calcaneus

I. Middle phalanges of 4 lateral toes

A. Flexion of PIP joints of 4 lateral toes

N. Medial plantar nerve (L4, 5)

P. Cannot palpate

Comparable to flexor digitorum superficialis in the forearm.

3. **Abductor Digiti Minimi**

O. Calcaneus

I. Base of proximal phalanx of little toe

A. Flexion, abduction of little toe at MP joint

N. Lateral plantar nerve (S1, 2)

P. Cannot palpate

Muscles of the Foot
Layer 2

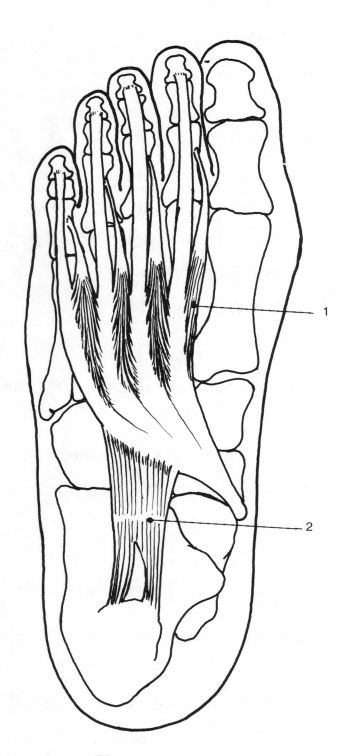

1. **Lumbricals**

   O. Tendons of flexor digitorum longus

   I. Extensor expansion to 4 lateral toes

   A. Flexion of MP joints
   Extension of DIP and PIP joints

   N. 1st: Medial plantar nerve (L4, 5)
   2nd, 3rd, 4th: Lateral plantar nerve (S1, 2)

   Comparable to lumbricals in the hand.

2. **Quadratus Plantae**

   O. Calcaneus

   I. Tendons of flexor digitorum longus

   A. Assists flexor digitorum longus in flexion of DIP joints

   N. Lateral plantar nerve (S1, 2)

   No counterpart in the hand.

Plantar

Muscles of the Foot Layer 3

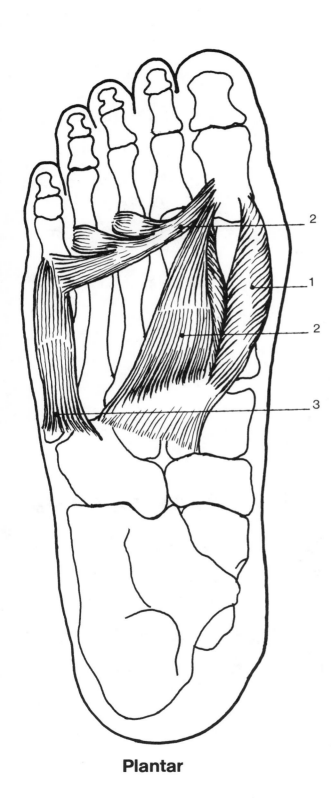

Plantar

1. **Flexor Hallucis Brevis**

   O. Base of metatarsal of great toe

   I. Base of proximal phalanx of great toe

   A. Flexion of MP joint (great toe)

   N. Medial plantar nerve (L4, 5, S1)

   Comparable to flexor pollicis brevis and has two heads.

2. **Adductor Hallucis**

   O. Oblique: Base of 2nd, 3rd metatarsals
   Transverse: 3rd, 4th, 5th MP joint capsules

   I. Base of proximal phalanx of great toe

   A. Adduction, flexion of great toe

   N. Lateral plantar nerve (S1, 2)

   Comparable to adductor pollicis in hand.

3. **Flexor Digiti Minimi Brevis**

   O. Cuboid and base of 5th metatarsal

   I. Base of proximal phalanx of 5th toe

   A. Flexion of MP joint

   N. Lateral plantar nerve (S1, 2)

   Comparable to flexor digiti minimi in hand.

Muscles of the Foot
Layer 4

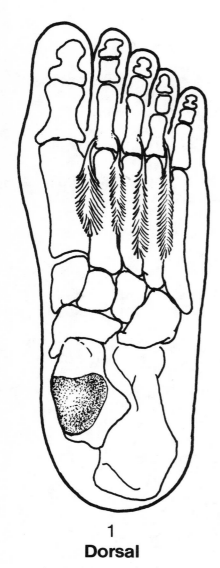

Dorsal

1. **Dorsal Interossei (4)**

   O. Adjacent metatarsals

   I. Extensor expansion of 2nd, 3rd and 4th toes

   A. Abduction of 2nd, 3rd, and 4th toes

   N. Lateral plantar nerve (S1, 2)

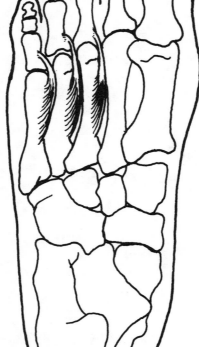

Plantar

2. **Plantar Interossei (3)**

   O. Medial side of 3rd, 4th and 5th metatarsals

   I. Extensor expansion to 3 lateral toes

   A. Adduction of 3 lateral toes

   N. Lateral plantar nerve (S1, 2)

Lateral (External) Rotators of the Hip

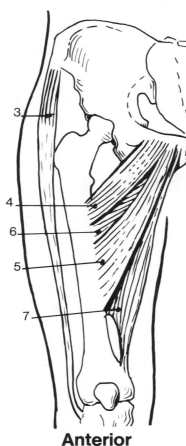

1. Piriforis
2. Gemellus superior
3. Obturator internus
4. Gemellus inferior
5. Obturator externus
6. Quadratus femoris
7. Gluteus maximus
8. Iliopsoas (not shown)

Posterior

Medial (Internal) Rotators of the Hip

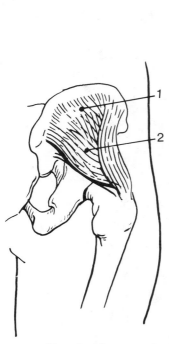

1. Gluteus medius
2. Gluteus minimus
3. *Tensor fasciae latae*
4. *Pectineus*
5. *Adductor longus*
6. *Adductor brevis*
7. *Adductor magnus (horizontal fibers)*

Posterior   **Anterior**

Extensors of the Hip

1. Gluteus maximus
2. Biceps femoris (long head)
3. Semitendinosus
4. Semimembranosus
5. *Adductor magnus (posterior portion)*

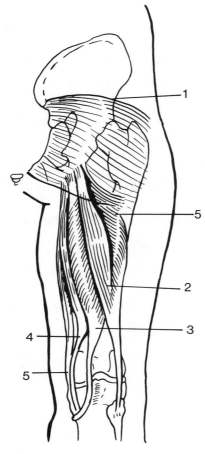

Posterior

Flexors of the Hip

1. Iliopsoas
2. Pectineus
3. *Tensor fasciae latae*
4. *Adductor brevis*
5. *Adductor longus*
6. *Adductor magnus (anterior portion)*
7. *Rectus femoris*
8. *Sartorius*

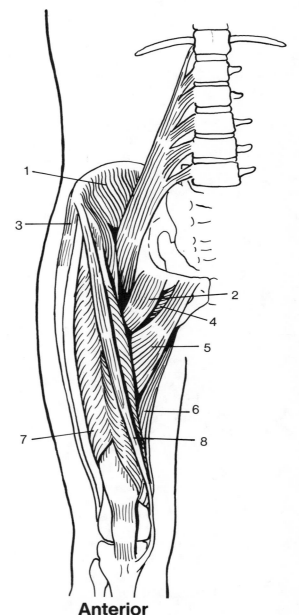

Anterior

Abductors of the Hip

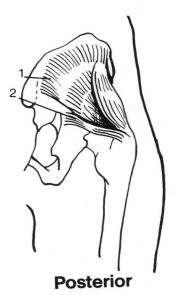

Posterior

1. Gluteus medius
2. Gluteus minimus
3. Iliopsoas
4. *Tensor fasciae latae*
5. *Sartorius*

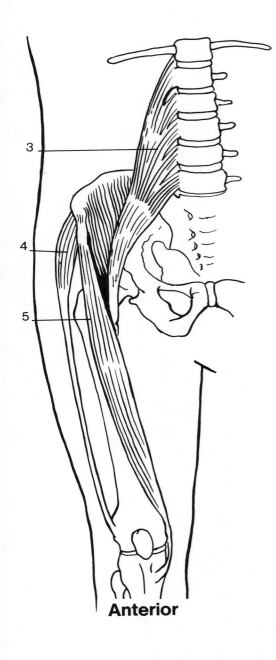

Anterior

Adductors of the Hip

1. Adductor brevis
2. Adductor longus
3. Adductor magnus
4. Gracilis
5. *Pectineus*

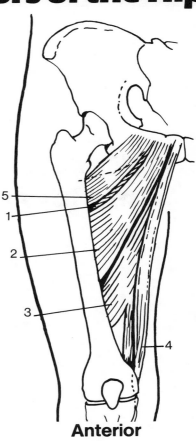

Anterior

Lateral (External) Rotator of the Knee

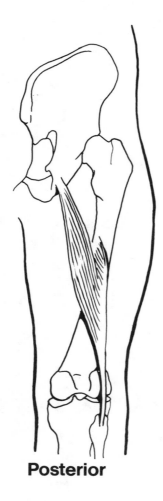

Biceps femoris

Posterior

Medial (Internal) Rotators of the Knee

1. Semitendinosus
2. Semimembranosus
3. Popliteus
4. *Gracilis*
5. *Sartorius*

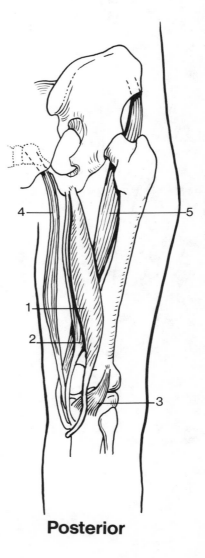

Posterior

Extensors of the Knee

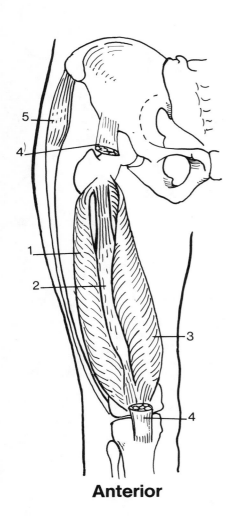

1. Vastus lateralis
2. Vastus intermedius
3. Vastus medialis
4. Rectus femoris
5. *Tensor fascia latae*

Anterior

Flexors of the Knee

1. Biceps femoris
2. Semitendinosus
3. Semimembranosus
4. *Sartorius*
5. *Gracilis*
6. *Gastrocnemius*
7. *Plantaris*
8. *Popliteus (not shown; deep to plantaris)*

Posterior

Dorsiflexors of the Ankle

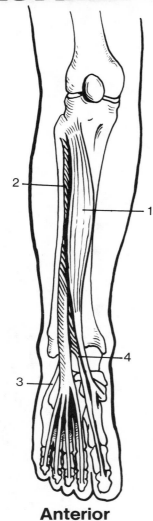

Anterior

1. Tibialis anterior
2. *Extensor digitorum longus*
3. *Peroneus tertius*
4. *Extensor hallucis longus*

Plantar Flexors of the Ankle

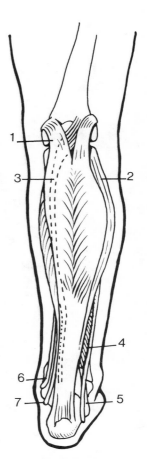

Posterior

1. Gastrocnemius
2. Soleus
3. *Plantaris*
4. *Peroneus longus*
5. *Peroneus brevis*
6. *Tibialis posterior*
7. *Flexor hallucis longus*
8. *Flexor digitorum longus* (not shown)

Invertors of the Foot

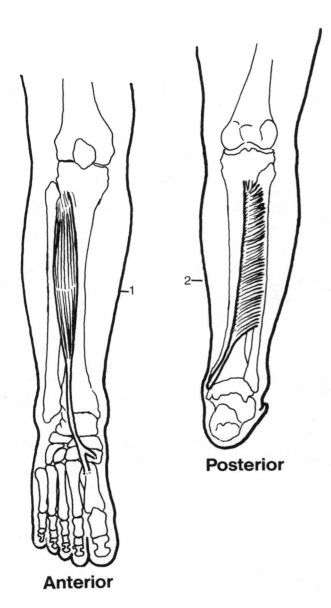

Anterior

Posterior

1. Tibialis anterior
2. Tibialis posterior

Evertors of the Foot

1. Peroneus tertius
2. Peroneus longus
3. Peroneus brevis

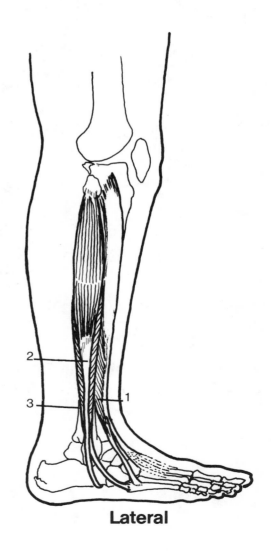

Lateral

Lumbosacral Plexus

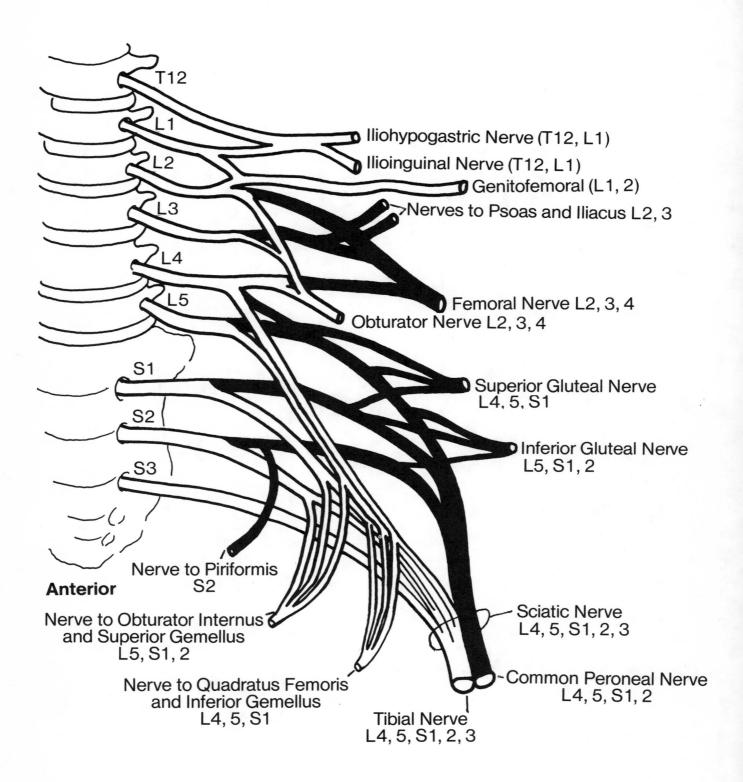

Obturator Nerve

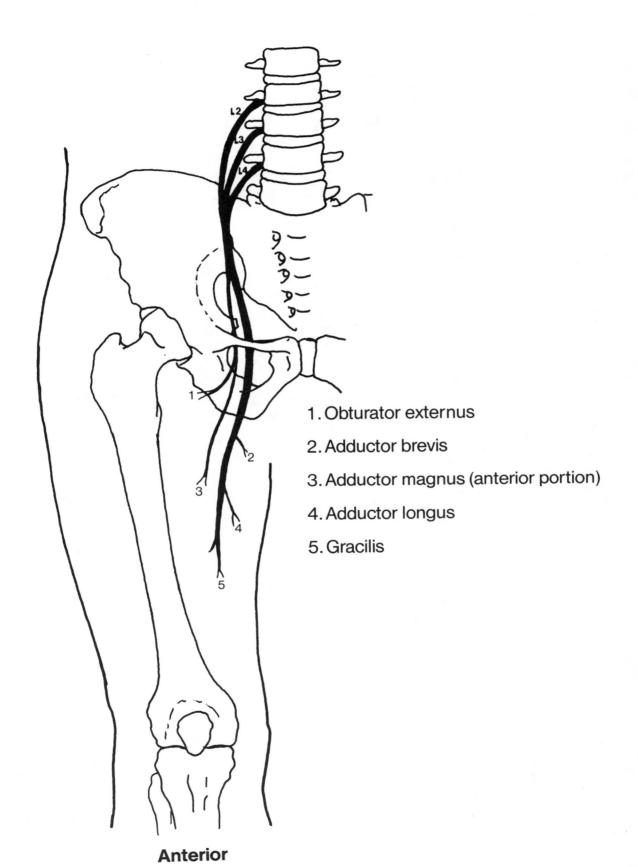

1. Obturator externus
2. Adductor brevis
3. Adductor magnus (anterior portion)
4. Adductor longus
5. Gracilis

Anterior

Femoral Nerve

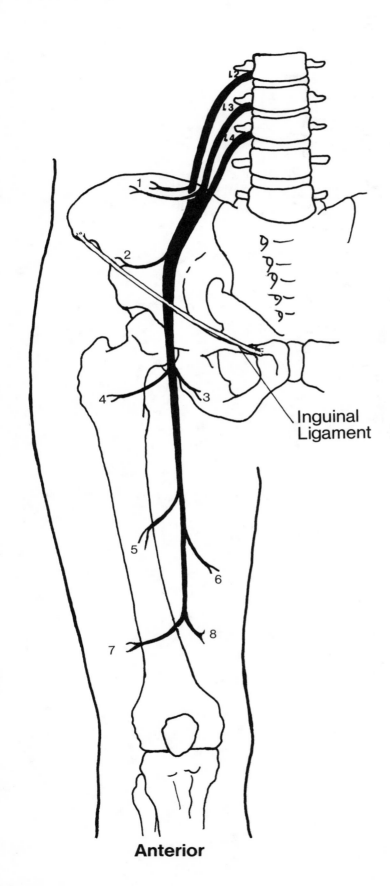

1. Psoas – L2, 3
2. Iliacus
3. Pectineus
4. Sartorius
5. Rectus femoris
6. Vastus medialis
7. Vastus lateralis
8. Vastus intermedius

Inguinal Ligament

Anterior

Sciatic Nerve

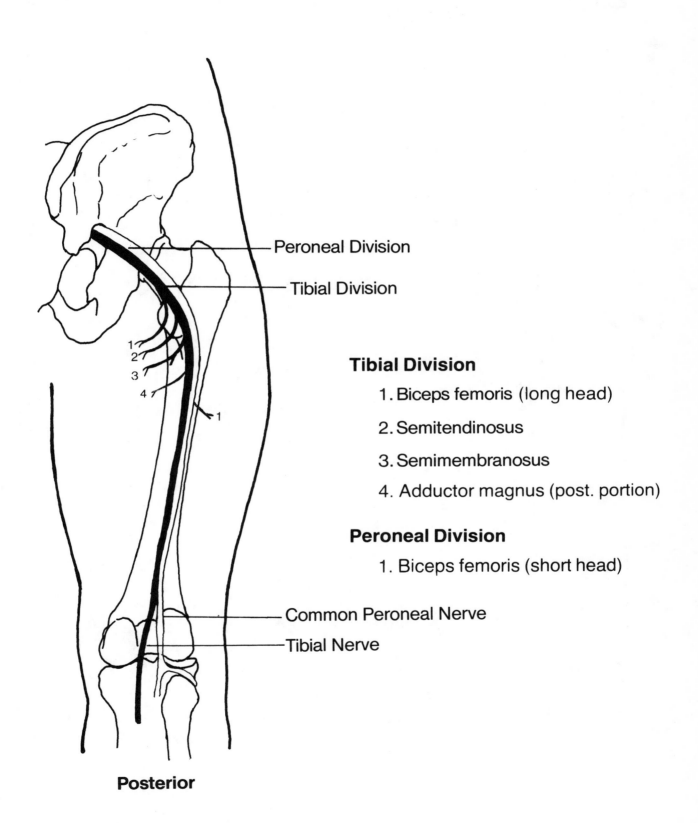

Tibial Division
1. Biceps femoris (long head)
2. Semitendinosus
3. Semimembranosus
4. Adductor magnus (post. portion)

Peroneal Division
1. Biceps femoris (short head)

Posterior

Peroneal Nerves
Common < Superficial / Deep

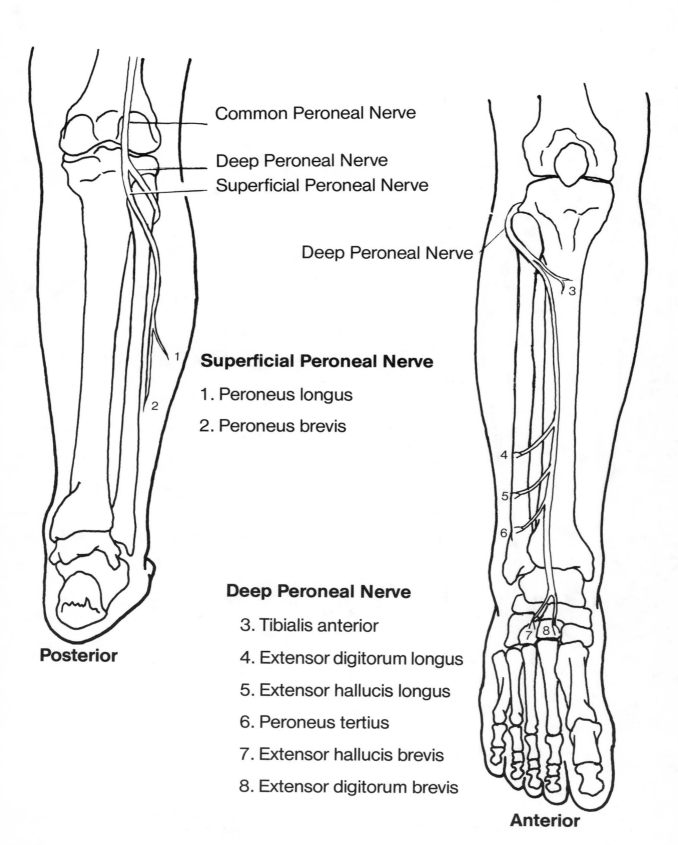

Common Peroneal Nerve
Deep Peroneal Nerve
Superficial Peroneal Nerve

Deep Peroneal Nerve

Superficial Peroneal Nerve
1. Peroneus longus
2. Peroneus brevis

Deep Peroneal Nerve
3. Tibialis anterior
4. Extensor digitorum longus
5. Extensor hallucis longus
6. Peroneus tertius
7. Extensor hallucis brevis
8. Extensor digitorum brevis

Posterior

Anterior

Tibial Nerve

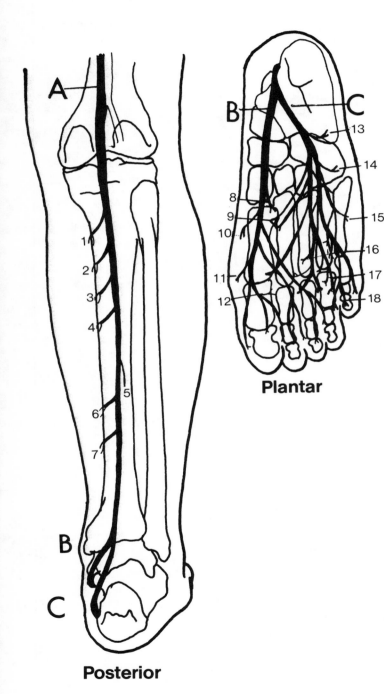

A. **Tibial Nerve**
1. Plantaris
2. Popliteus
3. Gastrocnemius
4. Soleus
5. Flexor digitorum longus
6. Flexor hallucis longus
7. Tibialis posterior

B. **Medial Plantar Nerve**
8. Flexor digitorum brevis
9. Abductor hallucis
10. Flexor hallucis brevis
11. 1st lumbrical

C. **Lateral Plantar Nerve**
12. Adductor hallucis
13. Quadratus plantae
14. Abductor digiti minimi
15. Flexor digiti minimi
16. Plantar interosseous
17. Dorsal interosseous
18. Lumbricals (lateral 3)

The tibial nerve innervates all muscles of the posterior lower leg.

Cutaneous Innervation of the Lower Limb

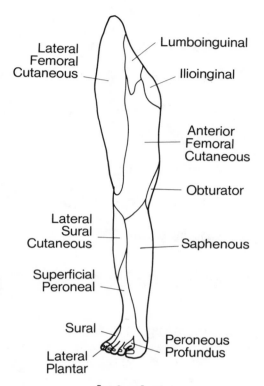
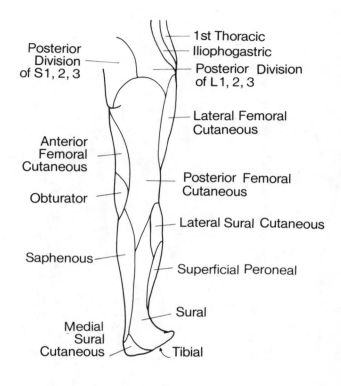

Anterior   **Posterior**

NERVE (Spinal Segments)	SOURCE
First Thoracic (T1)	T1
Posterior Divisions of L1, 2, 3	L1, 2, 3
Posterior Divisions of S1, 2, 3	S1, 2, 3
Iliohypogastric (L1)	Lumbar plexus
Ilioinguinal (L1)	Lumbar plexus
Lumboinguinal (L1, 2)	Genitofemoral nerve
Lateral femoral cutaneous (L2, 3)	Femoral nerve
Anterior femoral cutaneous (L2, 3)	Femoral nerve
Obturator (L2, 3, 4)	Terminal branches of obturator nerve
Saphenous (L2, 3, 4)	Obturator but travels in sheath with femoral nerve
Lateral sural cutaneous (L5, S1, 2)	Common peroneal nerve
Superficial peroneal (L4, 5, S1)	Common peroneal nerve
Peroneus profundus (L4, 5, S1)	Common peroneal nerve
Posterior femoral cutaneous S1, 2, 3	Tibial nerve
Medial sural cutaneous (S1, 2)	Tibial nerve
Sural (S1, 2)	Anastomoses of lateral and medial sural cutaneous nerves
Tibial (S1, 2)	Terminal branches of tibial nerve
Lateral plantar (S1, 2)	Tibial nerve

Extraocular Muscles

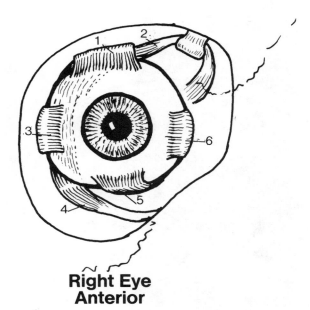

Right Eye Anterior

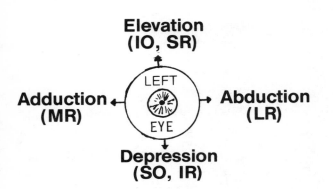

Right Eye Superior

1. **Superior Rectus (SR)**
2. **Superior Oblique (SO)**
3. **Lateral Rectus ((LR)**
4. **Inferior Oblique (IO)**
5. **Inferior Rectus (IR)**
6. **Medial Rectus (MR)**

O. Orbit

I. Eyeball

A. Superior rectus: elevation of eye
Superior oblique: depression of eye
Lateral rectus: abduction of eye
Inferior oblique: elevation of eye
Inferior rectus: depression of eye
Medial rectus: adduction of eye

Rotational movements of the eyeball around a sagittal plane (axis through pupil) also occur and are the result of combined actions of eye muscles. The movement described for a given muscle depends upon the starting position of the eye (e.g., so intorts or internally rotates the abducted eye).

N. Superior oblique: Cranial Nerve IV;
Lateral rectus: Cranial Nerve VI;
Other muscles: Cranial Nerve III

Elevation (IO, SR)
Adduction (MR) ← LEFT EYE → Abduction (LR)
Depression (SO, IR)

Innervation may be remembered as: "SO_4, LR_6, Others$_3$"

Muscles of Mastication, Lip and Jaw Closure

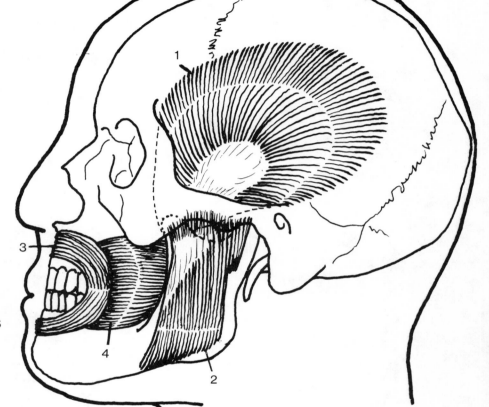

Lateral

1. Temporalis

O. Lateral surface of temporal bone

I. Mandible (coronoid process and ramus)

A. Closes jaw and retracts jaw

N. Cranial Nerve V (Trigiminal)

P. Lateral surface at temple

2. Masseter

O. Zygomatic arch

I. Mandible (lateral surface of ramus)

A. Closes jaw and assists protraction
Unilaterally: lateral jaw motion to same side

N. Cranial Nerve V (Trigiminal)

P. Lateral mandible in area of back molars

3. Orbicularis Oris

O. Maxilla, mandible, lips, buccinator

I. Mucous membranes, muscles inserting into lip

A. Lip closure

N. Cranial Nerve VII (Facial)

P. Around lips

4. Buccinator

O. Maxilla, mandible

I. Lips

A. Maintains cheeks near teeth (and thereby food in position for chewing)

N. Cranial Nerve VII (Facial)

P. Cheeks

Muscles of Mastication, Lip and Jaw Closure (continued)

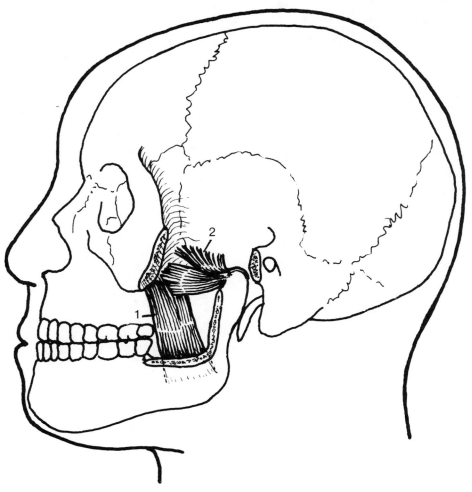

Lateral

1. Medial Pterygoid

O. Pterygoid plate (medial surface)

I. Mandible (medial surface of ramus)

A. Closes jaw and assists protraction
Unilaterally: lateral jaw motion to opposite side

N. Cranial Nerve V (Trigiminal)

P. Cannot palpate

2. Lateral Pterygoid

O. Pterygoid plate (lateral surface)

I. Mandible Temporomandibular joint capsule

A. Protracts jaw
Unilaterally: lateral jaw motion to opposite side

N. Cranial Nerve V (Trigiminal)

P. Cannot palpate

Muscles of Deglutition

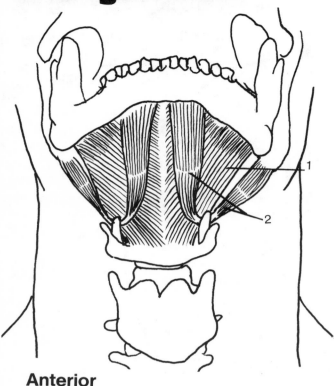

Anterior

Suprahyoids
1. Mylohyoids
2. Digastrics (anterior and posterior bellies)
3. Geniohyoids
   Stylohyoids (not shown – small muscle running with posterior belly of digastric)

Infrahyoids
4. Thyrohyoid
5. Sternothyroid
6. Sternohyoid
7. Omohyoid

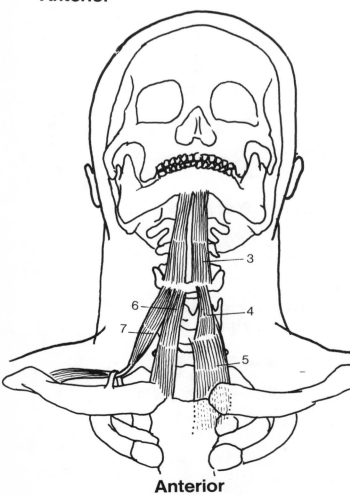

Anterior

Suprahyoids
O. Mandible

I. Hyoid bone

A. Raises hyoid bone in swallowing (when mandible is stable) **or** opens the jaw (when hyoid bone is stable)

N. Mylohoid – Cranial Nerve V
   Digastrics –
     anterior belly – Cranial Nerve V
     posterior belly – Cranial Nerve VII
   Geniohyoid – C1 via Cranial Nerve XII
   Stylohyoid – Cranial Nerve VII

P. Mylohyoid can be felt with and underneath the tongue during swallowing

Infrahyoids
O. Thyroid cartilage, manubrium, superior border of scapula

I. Hyoid bone

A. Stabilizes hyoid bone in swallowing by pulling it downward

N. C1, 2, 3

P. Difficult to differentiate from other anterior muscles of the neck

Sternocleidomastoid

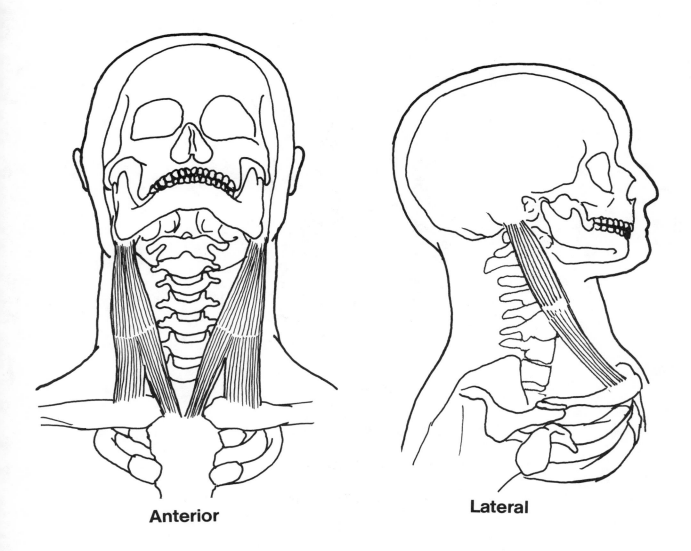

Anterior **Lateral**

- O. Manubrium of sternum
  Medial clavicle

- I. Mastoid process

- A. Bilaterally – flexion of neck
  Unilaterally – lateral flexion, rotation of head to opposite side

- N. Accessory nerve (or Cranial Nerve XI) (C2, 3)

- P. Anterior-lateral neck, diagonally between its origin and insertion

Sternocleidomastoid is named for its attachments.

Scalenes (Anterior, Medius, Posterior)

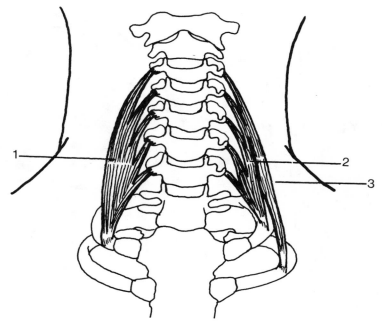

Anterior

1. Scalenus Anterior
2. Scalenus Medius
3. Scalenus Posterior

O. Transverse processes of cervical vertebrae

I. First 2 ribs (anterior and medius to 1st rib; posterior to 2nd rib)

A. Bilaterally – raise first 2 ribs during forced inspiration **or** assist neck flexion
Unilaterally – assist in lateral flexion to same side

N. Branches of Cervical Nerves 6, 7, 8

P. Cannot palpate

Scalenes is Greek for "uneven," which describes the shape of these three muscles. The brachial plexus passes between the anterior scalenus and middle scalenus.

Rectus Abdominis

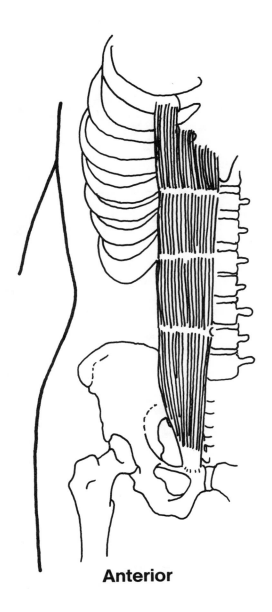

Anterior

O. Costal cartilages 5, 6, 7

I. Pubis

A. Flexion of trunk
   Compression of abdominal contents

N. 7-12 intercostal nerves

P. Anterior-medial surface of abdomen, on either side of umbilicus from sternum to pubis

Rectus abdominis contracts strongly when doing sit-ups or lifting both legs several inches from the floor in a supine position.

Some sources consider O. and I. in terms of kinesiological muscle action and thus, for the rectus abdominus, the O. is the pubis and the I. is the ribs. However, in this manual, the anatomical proximal-distal determination of O. and I. has been used throughout.

External Oblique

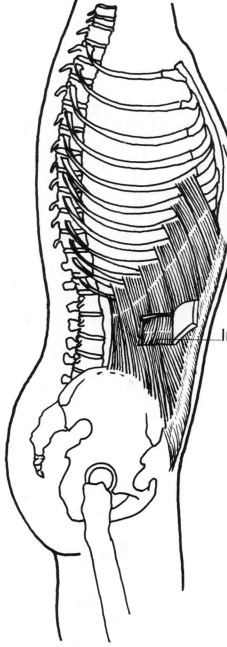

Lateral

O. Lower 8 ribs

I. Abdominal aponeurosis and iliac crest

A. Bilaterally – flexion of trunk, compression of abdominal contents
Unilaterally – lateral flexion, rotation of trunk to opposite side

N. 8-12 intercostal nerves
L1 (L1 = iliohypogastric nerve and ilioinguinal nerve)

P. Lateral sides of abdomen

Internal oblique

If you position your hands as if you were reaching into your pants pocket, your fingers will assume the direction of these fibers (obliquely downward and medialward). Its origin interdigitates with serratus anterior, and it is the most superficial side muscle.

Internal Oblique

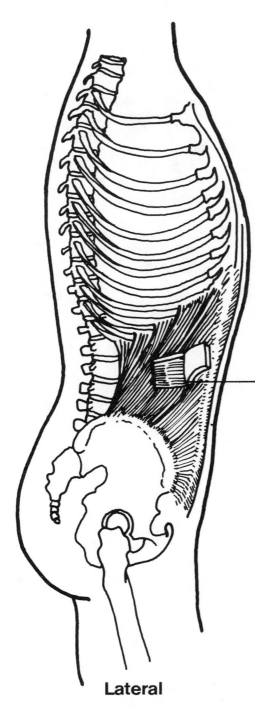

Lateral

O. Inquinal ligament and anterior iliac crest

I. Costal cartilages of last 4 ribs
   Abdominal aponeurosis

A. Bilaterally – flexion of spine, compression of abdominal contents
   Unilaterally – lateral flexion, rotation of trunk to same side

N. 8-12 intercostal nerves
   L1 (L1 = iliohypogastric nerve and ilioinguinal nerve)

P. Cannot palpate

— Transversus abdominis

If you cross your arms over your abdomen with fingertips on the anterior-superior iliac spines your fingertips will assume the direction of these fibers (obliquely upward and medialward).

Transverse Abdominis

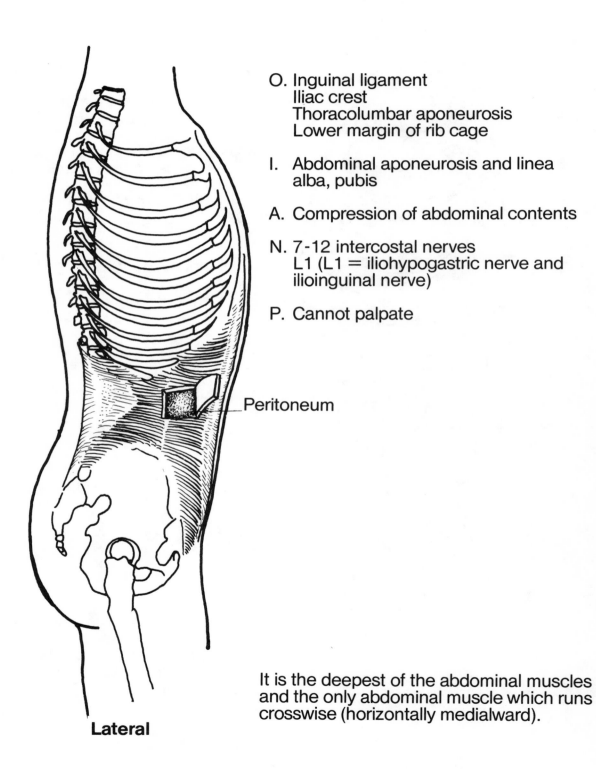

O. Inguinal ligament
Iliac crest
Thoracolumbar aponeurosis
Lower margin of rib cage

I. Abdominal aponeurosis and linea alba, pubis

A. Compression of abdominal contents

N. 7-12 intercostal nerves
L1 (L1 = iliohypogastric nerve and ilioinguinal nerve)

P. Cannot palpate

Peritoneum

Lateral

It is the deepest of the abdominal muscles and the only abdominal muscle which runs crosswise (horizontally medialward).

Intercostals

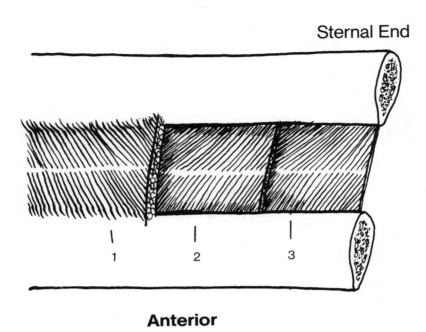

Sternal End

Anterior

1. **External Intercostals**
   (exterior surface of ribs)

2. **Internal Intercoastals**
   (interior surface of ribs)

3. **Innermost Intercostals**
   (deep to internal intercostals)

O. Between adjacent ribs

I. Between adjacent ribs

A. Elevate ribs in inspiration
   Maintains intercostal spaces

N. Intercostal nerves

P. External intercostals are barely palpable between ribs

These muscles are the thoracic continuation of the external and internal obliques.

Intercostal nerves are the anterior rami of the thoracic spinal nerves.

Diaphragm

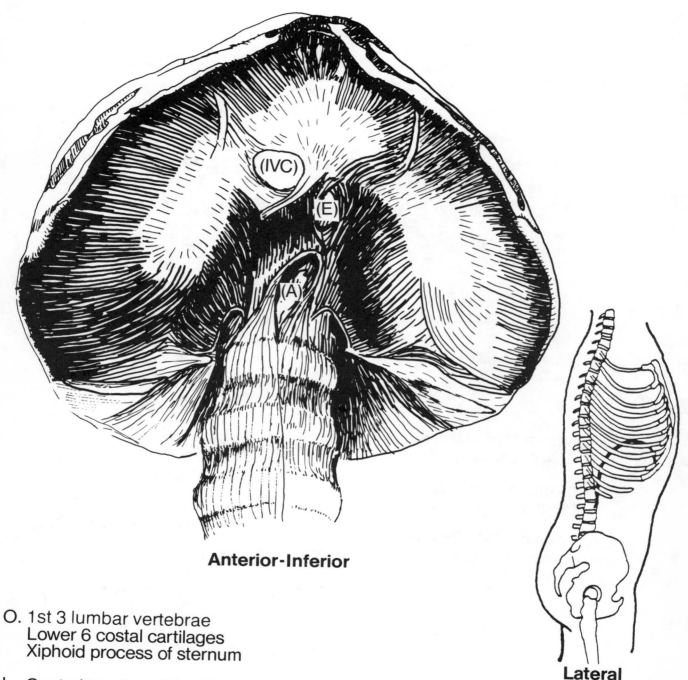

Anterior-Inferior

Lateral

O. 1st 3 lumbar vertebrae
   Lower 6 costal cartilages
   Xiphoid process of sternum

I. Central tendon of the diaphragm
   (clover leaf shaped aponeurosis)

A. Flattens central tendon and thus increases vertical diameter of thoracic cavity in inspiration

N. Phrenic nerve (C3, 4, 5)

P. Cannot palpate

The diaphragm contracts down in inspiration and relaxes in expiration. It separates the thoracic cavity from the abdominal cavity and has three openings to allow passage of the esophagus, inferior vena cava and aorta, labeled (E), (IVC) and (A) in the above drawing. The drawing in the lateral view shows the diaphragm in situ.

Serratus Posterior Superior
Serratus Posterior Inferior

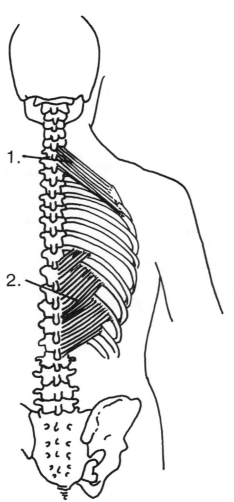

Posterior

1. **Serratus Posterior Superior**

   O. Caudal ligamentum nuchae
   Spinous processes of C7 - T2
   Supraspinal ligament

   I. Cranial borders of 2, 3, 4, 5 ribs

   A. Raises ribs to increase thoracic cavity

   N. 1-4 intercostal nerves

   P. Cannot palpate

2. **Serratus Posterior Inferior**

   O. Spinous processes of T11, 12 and L1-3
   Supraspinal ligament

   I. Inferior borders of last 4 ribs

   A. Draws ribs outward and downward counteracting the inward pull of the diaphragm

   N. 9 - 12 intercostal nerves

These 2 thin muscles on the posterior thorax are considered respiratory muscles. Serratus posterior superior lies beneath the levator scapula and the upper trapezius; serratus posterior inferior lies beneath the latissimus dorsi.

Quadratus Lumborum

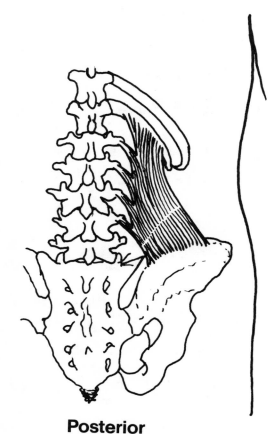

Posterior

O. Posterior iliac crest

I. 12th rib
   Transverse processes of lumbar vertebrae

A. Lateral flexion of trunk **or** raises hip

N. Branches of T12, L1 nerves

P. Cannot palpate

This muscle can be remembered for its action as a "hip hiker".

Splenius Capitis
Splenius Cervicis

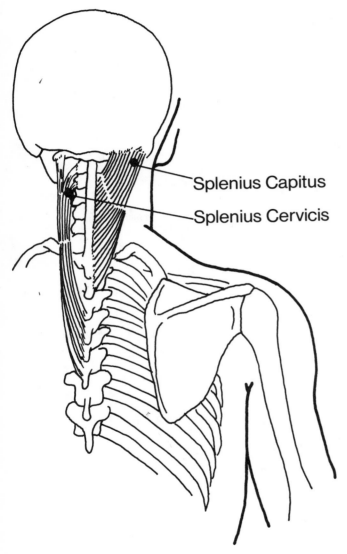

Posterior

O. Splenius capitus:
   Ligamentum nuchae
   Lower cervical vertebra (C7)
   Upper thoracic vertebrae (T1, 2, 3)

   Splenius cervicis:
   Upper thoracic vertebrae (T3 - 6)

I. Splenius capitis:
   Mastoid process and occipital bone

   Splenius cervicis:
   Upper cervical vertebrae (transverse processes of C1, 2, 3)

A. Bilaterally – extension of neck
   Unilaterally – rotation of head to same side

N. Posterior branches of Cervical Nerves

P. With difficulty, on posterior neck between trapezius and sternocleidomastoid above levator scapula

The splenius muscles are names for the Latin term for bandage because they appear to wrap around deeper neck muscles. Splenius capitis inserts deep to sternocleidomastoid.

Erector Spinae

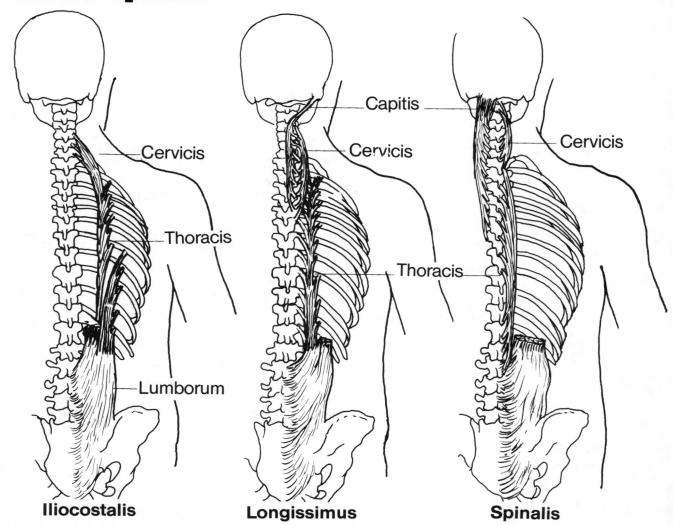

Iliocostalis   **Longissimus**   **Spinalis**

O. Iliocostalis (lateral layer): Thoracolumbar aponeurosis, posterior ribs

Longissimus (middle layer): thoracolumbar aponeurosis, lumbar and thoracic transverse processes

Spinalis (medial layer): ligamentum nuchae, cervical and thoracic spinous processes

I. Iliocostalis: Posterior ribs, cervical transverse processes

Longissimus: cervical and thoracic transverse processes, mastoid process

Spinalis: cervical and thoracic spinous processes, occipital bone

A. Bilaterally – extension of spine
Unilaterally – lateral flexion of spine

N. Posterior branches of spinal nerves

P. Cannot palpate

Erector spinae are covered by thoracolumbar and nuchael fascia and lie superficial to transversospinalis. The lateral layer (iliocostalis) has attachments throughout the spinal column while the middle (longissimus) and medial (spinalis) layers attach to the skull and cervical and thoracic vertebrae. The insertion of longissimus to the skull is found deep to spenius capitis. The skull portion of spinalis is actually inseparable from the skull portion of semispinalis which is described with transversospinalis.

Transversospinalis – Semispinalis and Multifidus

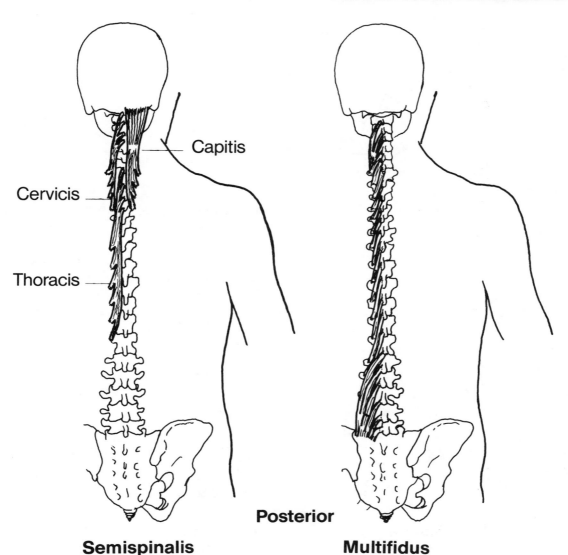

Semispinalis   **Posterior**   **Multifidus**

O. Semispinalis: cervical and thoracic transverse processes

    Multifidus: sacrum, posterior superior iliac spine, transverse processes all vertebrae

I. Semispinalis: cervical and thoracic spinous processes, occipital bone (spans 3 to 6 vertebrae)

    Multifidus: spinous processes of all vertebrae, inserting 2 to 4 vertebrae above origin

A. Bilaterally – extension of spine
    Unilaterally – rotation of opposite side

N. Posterior branches of spinal nerves

P. Cannot palpate

Transverospinalis muscles lie deep to erector spinae and are found in order from superficial to deep: semispinalis; multifidus; and rotatores, interspinales and intertransversarii. Deeper muscles span one or two vertebrae while superficial layers have longer spans.

Transversospinalis – Rotatores, Interspinales and Intertransversarii

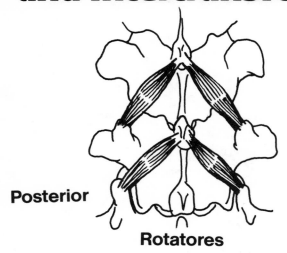
Posterior
Rotatores

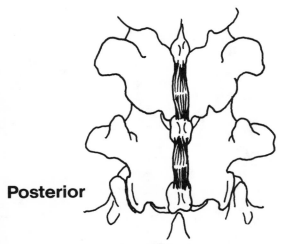
Posterior
Interspinales

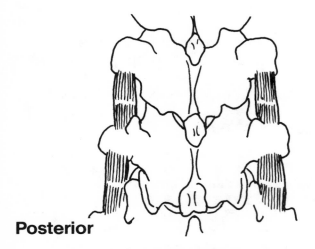
Posterior
Intertransversarii

O. Rotatores: transverse processes of all vertebrae

Interspinalis: spinous processes of cervical, lumbar and T1-2, T11-12 vertebrae

Intertransversarii: transverse processes of cervical, lumbar and T10-12 vertebrae

I. Rotatores: spinous processes of all vertebrae, inserting on vertebra directly above origin

Interspinalis: spinous processes of cervical, lumbar and T1-2, T11-12 vertebrae, each inserting on vertebra directly above origin

Intertransversarii: transverse processes cervical, lumbar and T10-12, inserting on vertebra directly above origin

A. Rotatores: extension of spine and rotation to opposite side

Interspinalis: extension of spine

Intertransversarii: lateral flexion of spine

N. Posterior branches of spinal nerves

P. Cannot palpate

REFERENCES

1. Basmajian JV: *Primary Anatomy*, 8th Edition, Baltimore, Williams and Wilkins Co, 1982.
2. Basmajian JV (Ed): *Grant's Method of Anatomy by Regions Descriptive and Deductive*, Baltimore, Williams and Wilkins Co, 1980.
3. Basmajian JV: *Surface Anatomy*, Baltimore, Williams and Wilkins, 1977.
4. Brunnstrom S: *Clinical Kinesiology*, 4th Edition, Philadelphia, FA Davis Co, 1983.
5. Chusid JG: *Correlative Neuroanatomy and Functional Neurology*, 18th Edition, Los Altos, California, Lange Medical Publications, 1982.
6. Crouch JE: *Functional Human Anatomy*, Philadelphia, Lea and Febiger, 1965.
7. Daniels L, Williams M, and Worthingham C: *Muscle Testing: Techniques of Manual Examination*, 4th Edition, Philadelphia, WB Saunders Co, 1980.
8. Ferner H (Editor): *Edward Perkopf Atlas of Topographical and Applied Human Anatomy*, Philadelphia, Lea and Febiger, 1973.
9. Goss CM (Editor): *Gray's Anatomy*, 29th Edition, Philadelphia, Lea and Febiger, 1973.
10. Grant CJB: *An Atlas of Anatomy*, Baltimore, Williams and Wilkins Co, 1962.
11. Jacob SW, Francone CA: *Structure and Function in Man*, Philadelphia, WB Saunders Co, 1982.
12. Kaplan EB: *Functional and Surgical Anatomy of the Hand*, Philadelphia, JP Lippincott Co, 1965.
13. Kendall HO, Kendall FP, and Wadsworth GE: *Muscles – Testing and Function*, 3rd Edition, Baltimore, Williams and Wilkins Co, 1983.
14. Killingsworth A: *Basic Physical Disability Procedures*, Oakland, California, Cal-Syl Press, 1976.
15. Lockhart RD, Hamilton GF, and Fyfe FW: *Anatomy of the Human Body*, Philadelphia, JB Lippincott Co, 1960.
16. Medical Research Council: *Aids to the Examination of the Peripheral Nervous System*, London, Her Majesty's Stationery Office, 1976.
17. Miller MA, Leavell LC: *Anatomy and Physiology*, New York, MacMillan Publishing Co, 1972.
18. Montgomery RL: *Basic Anatomy for the Allied Health Professions*, Baltimore, Urkan & Schwartanberg, 1980.
19. Moore KL: *Clinically Oriented Anatomy*, Baltimore, Williams & Wilkins Co, 1980.
20. Trombly CA, *Occupational Therapy for Physical Dysfunction*, Baltimore, Williams and Wilkins Co, 1983.
21. Walton JH: *Surgical Anatomy of the Hand*, Summit, New Jersey, Ciba Pharmaceutical Co, 1969.
22. Walwich R, Williams PL (Editors), *Gray's Anatomy*, 35th British Edition, Philadelphia, WB Saunders Co, 1973.
23. Wells KF: *Kinesiology: The Scientific Basis of Human Motion*, 5th Edition, Philadelphia, WB Saunders Co, 1971.
24. Woodburn RT: *Essentials of Human Anatomy*, New York, Oxford University Press, 1965.
25. Zuckerman, Sir Solly: *A New System of Anatomy*, London, Oxford University Press, 1961.

INDEX

A

Abductor digiti minimi, 52, 115
Abductor hallucis, 115
Abductor pollicis brevis, 46
Abductor pollicis longus, 36
Abductors, 65, 66, 70, 71, 121
Adductor brevis, 97
Adductor hallucis, 117
Adductor longus, 97
Adductor magnus, 98
Adductor pollicis, 47
Adductors, 65, 66, 70, 71, 121
Ankle, 124, 125
Anconeus, 27
Arm, 61-66
Axillary nerve (circumflex), 74

B

Biceps brachii, 25
Biceps femoris, 100
Bones, 1-10; 80-85
Brachialis, 26
Brachial plexus, 73
Brachioradialis, 28
Buccinator, 134

C

Carpals, 8
Circumflex nerve (axillary), 74
Clavicle, 4
Coracobrachialis, 17
Cutaneous innervation of lower limb, 132
Cutaneous innervation of upper limb, 79

D

Deglutition, 136
Deltoids, 16
Depressors of scapula, 58
Diaphragm, 144
Digastric, 136
Digits, 71, 72
Dorsal (extensor) expansion, 56
Dorsal interossei, 53, 118
Dorsiflexor of ankle, 124
Downward rotation of scapula, 60

E

Elbow flexion and extension, 67
Elevators of scapula, 58
Erector spinae, 148
Extensor carpi radialis brevis, 30
Extensor carpi radialis longus, 29
Extensor carpi ulnaris, 31
Extensor digiti minimi, 33
Extensor digitorum, 32
Extensor digitorum brevis, 110
Extensor digitorum longus, 111
Extensor hallucis longus, 109
Extensor indicis, 34
Extensor pollicis brevis, 37
Extensor pollicis longus, 38
Extensors, 64, 67, 69, 72, 120, 123
External oblique, 140
Extraocular muscles, 133
Evertors of foot, 125
Eye muscles, 133

F

Femoral nerve, 128
Femur, 81
Fibula, 82
Flexor carpi radialis, 42
Flexor carpi ulnaris, 40
Flexor digiti minimi, 49
Flexor digiti minimi brevis, 117
Flexor digitorum brevis, 115
Flexor digitorum longus, 106
Flexor digitorum profundus, 44
Flexor digitorum superficialis, 43
Flexor hallucis brevis, 117
Flexor hallucis longus, 107
Flexor pollicis brevis, 48
Flexor pollicis longus, 45
Flexors, 63, 67, 69, 72, 120, 123
Foot, 83, 115-118
Forearm, 68

Continued on next page.

G
Gastrocnemius, 102
Gemellus inferior, 90
Gemellus superior, 90
Geniohyoid, 136
Gluteus maximus, 86
Gluteus medius, 87
Gluteus minimus, 88
Gracilis, 99

H
Hamstrings, 100, 101
Hand, 8; 56-57; 69-72
Hip, 90; 119-122
Humerus, 6; 61-66

I
Iliacus 91
Iliocostalis, 148
Iliopsoas, 91
Ilium, 80
Inferior oblique, 133
Inferior rectus, 133
Infrahyoid, 136
Infraspinatus, 19
Intercostals, 143
Internal oblique, 141
Interossei, 53, 54, 118
Interspinalis, 150
Intertransversarii, 150
Intrinsics of foot, 115-118
Intrinsics of hand, 71, 72
Invertors of foot, 125
Ischium, 80

K
Knee, 122, 123

L
Lateral pterygoid, 135
Lateral rectus, 133
Lateral rotators, 62, 90, 119, 122
Latissimus dorsi, 12
Levator scapula, 14
Longissimus, 148
Lumbosacral plexus, 126
Lumbricals, 55, 116

M
Masseter, 134
Mastication, 134, 135
Medial pterygoid, 135
Medial rectus, 133
Medial rotators, 61, 119, 122
Median nerve, 77
Multifidus, 149
Musculocutaneous nerve, 75
Mylohyoid, 136

N
Nerves, 73-79; 126-132

O
Orbicularis oris, 134
Obturator externus, 90
Obturator internus, 90
Obturator nerve, 127
Omohyoid, 136
Opponens digit minimi, 51
Opponens pollicis, 50
Opposition, 72

P
Palmar interossei, 54
Palmaris longus, 41
Palpation
   bones: 9-10, 84-85
    muscles: see individual muscles
Pectineus, 96
Pectoralis major, 22
Pectoralis minor, 23
Pelvis, 80
Peroneal nerves, 130
Peroneus brevis, 113
Peroneus longus, 114
Peroneus tertius, 112
Piriformis, 90
Plantarflexors of ankle, 124
Plantar interossei, 118
Plantaris, 104
Popliteus, 104
Pronator quadratus, 39
Pronator teres, 39
Pronators of forearm, 68
Protractors of scapula, 59
Psoas major and minor, 91

Continued on next page.

Q
Quadratus femoris, 90
Quadratus lumborum, 146
Quadratus plantae, 116
Quadriceps femoris, 93-95

R
Radial nerve, 76
Radius, 7, 68
Rectus abdominis, 139
Rectus femoris, 93, 94
Retractors of scapula, 59
Rhomboids, 15
Ribs, 4
Rotatores, 150

S
Sacrum, 80
Sartorius, 92
Scalenes anterior, medius, posterior, 138
Scapula, 5, 58-60
Sciatic nerve, 129
Semimembranosus, 100, 101
Semispinalis, 149
Semitendinosus, 100, 101
Serratus anterior, 24
Serratus posterior inferior, 145
Serratus posterior superior, 145
Skeleton, 1
Skull, 2
Soleus, 103
Spinalis, 148
Splenius capitis, cervicis, 147
Sternocleidomastoid, 137
Sternohyoid, 136
Sternothyroid, 136
Sternum, 4
Stylohyoid, 136
Subclavius, 23
Subscapularis, 21
Superior oblique, 133
Superior rectus, 133
Supinator, 35
Supinators of forearm, 68
Suprahyoids, 136
Supraspinatus, 18

T
Temporalis, 134
Tendon
   digits, 56
   wrist, 56, 57
Tensor fascia latae, 89
Teres major, 13
Teres minor, 20
Thumb, 71, 72
Thyrohyoid, 136
Tibia, 82
Tibialis anterior, 108
Tibialis posterior, 105
Tibial nerve, 131
Transversospinalis, 149, 150
Transverse abdominis, 142
Trapezius, 11
Triceps brachii, 27
Trunk, 139

U
Ulna, 7, 68
Ulnar nerve, 78
Upward rotation of scapula, 60

V
Vertebral column, 3
Vastus intermedius, 93, 95
Vastus lateralis, 93, 95
Vastus medialis, 93, 95

W
Wrist, 69-71
Wrist tendon compartments, 55, 56

of Books
Ordered: **Pricing** **Discounts**
1 - $19.95 ea: (None)
2 - 50 - $17.95 ea. (10%)
51 - 150 - $15.95 ea. (20%)
151 or more - $13.95 ea. (30%)

Orders for single copies must be PREPAID.
rs from institutions must be prepaid or mpanied by a signed purchase order. ject to credit approval.

Shipping and Handling
All orders shipped U.P.S.
Single Copy Orders
st be prepaid plus $2.75 shipping.
Multiple Copy Orders
e invoiced to include shipping and ling charges.

MEGABOOKS
300 N.W. 23rd Ave., Suite 192
P.O. Box 1702
Gainesville, FL 32606-9990
(904) 371-6342

Illustrated Essentials of Musculoskeletal Anatomy
ORDER FORM

Customer Purchase Order Number _____

Ship to: Name _____ Title _____
Discipline/Profession _____
Institution _____
Address _____
City _____ State _____ Zip _____
Telephone (_____) _____

_____ books @ $ _____ = _____
Florida Res. add 6% Sales Tax _____

☐ MasterCard ☐ VISA   Tax exempt No. _____
☐ Choice   Shipping and Handling Charges _____

Exp. Date _____   Total Due _____
Card No. _____

NEW

Signature _____

Send me information on:
☐ **Study cards** ☐ **Transparencies**

---

of Books
Ordered: **Pricing** **Discounts**
1 - $19.95 ea: (None)
2 - 50 - $17.95 ea. (10%)
51 - 150 - $15.95 ea. (20%)
151 or more - $13.95 ea. (30%)

Orders for single copies must be PREPAID.
rs from institutions must be prepaid or mpanied by a signed purchase order. ject to credit approval.

Shipping and Handling
All orders shipped U.P.S.
Single Copy Orders
st be prepaid plus $2.75 shipping.
Multiple Copy Orders
e invoiced to include shipping and ling charges.

MEGABOOKS
300 N.W. 23rd Ave., Suite 192
P.O. Box 1702
Gainesville, FL 32606-9990
(904) 371-6342

Illustrated Essentials of Musculoskeletal Anatomy
ORDER FORM

Customer Purchase Order Number _____

Ship to: Name _____ Title _____
Discipline/Profession _____
Institution _____
Address _____
City _____ State _____ Zip _____
Telephone (_____) _____

_____ books @ $ _____ = _____
Florida Res. add 6% Sales Tax _____

☐ MasterCard ☐ VISA   Tax exempt No. _____
☐ Choice   Shipping and Handling Charges _____

Exp. Date _____   Total Due _____
Card No. _____

NEW

Signature _____

Send me information on:
☐ **Study cards** ☐ **Transparencies**